Springer Undergraduate Matl

Springer

London
Berlin
Heidelberg
New York
Barcelona
Hong Kong
Milan
Paris
Singapore
Tokyo

Advisory Board

Other books in this series

T.S. Blyth and E.F. Robertson

Further Linear Algebra

 Springer

Professor T.S. Blyth
Professor E.F. Robertson

School of Mathematics and Statistics, University of St Andrews, North Haugh,
St Andrews, KY16 9SS, UK

Cover illustration elements reproduced by kind permission of:

Aptech Systems, Inc., Publishers of the GAUSS Mathematical and Statistical System, 23804 S.E. Kent-Kangley Road, Maple Valley, WA 98038, USA. Tel: (206) 432 - 7855 Fax (206) 432 - 7832 email: info@aptech.com URL: www.aptech.com

American Statistical Association: Chance Vol 8 No 1, 1995 article by KS and KW Heiner 'Tree Rings of the Northern Shawangunks' page 32 fig 2

Springer-Verlag: Mathematica in Education and Research Vol 4 Issue 3 1995 article by Roman E Maeder, Beatrice Amrhein and Oliver Gloor 'Illustrated Mathematics: Visualization of Mathematical Objects' page 9 fig 11, originally published as a CD ROM 'Illustrated Mathematics' by TELOS: ISBN 0-387-14222-3, German edition by Birkhauser: ISBN 3-7643-5100-4.

Mathematica in Education and Research Vol 4 Issue 3 1995 article by Richard J Gaylord and Kazume Nishidate 'Traffic Engineering with Cellular Automata' page 35 fig 2. Mathematica in Education and Research Vol 5 Issue 2 1996 article by Michael Trott 'The Implicitization of a Trefoil Knot' page 14.

Mathematica in Education and Research Vol 5 Issue 2 1996 article by Lee de Cola 'Coins, Trees, Bars and Bells: Simulation of the Binomial Process' page 19 fig 3. Mathematica in Education and Research Vol 5 Issue 2 1996 article by Richard Gaylord and Kazume Nishidate 'Contagious Spreading' page 33 fig 1. Mathematica in Education and Research Vol 5 Issue 2 1996 article by Joe Buhler and Stan Wagon 'Secrets of the Madelung Constant' page 50 fig 1.

British Library Cataloguing in Publication Data
Blyth, T.S. (Thomas Scott)
   Further linear algebra. - (Springer undergraduate
   mathematics series)
   1. Algebra, Linear
   I. Title II. Robertson, E.F. (Edmund F)
   512.5
ISBN 1852334258

Library of Congress Cataloging-in-Publication Data
Blyth, T.S. (Thomas Scott)
   Further linear algebra / T.S. Blyth and E.F. Robertson.
     p.   cm. -- (Springer undergraduate mathematics series, ISSN 1615-2085)
   Includes index.
   ISBN 1-85233-425-8 (acid-free paper)
   1. Algebras, Linear. I. Robertson, E.F. II. Title. III. Series.
QA184.2.B59   2002
512'.55—dc21   2001042987

Springer Undergraduate Mathematics Series ISSN 1615-2085
ISBN 1-85233-425-8 Springer-Verlag London Berlin Heidelberg
a member of BertelsmannSpringer Science+Business Media GmbH
http://www.springer.co.uk

Typesetting: Camera ready by T.S.B.
Printed and bound at the Athenæum Press Ltd., Gateshead, Tyne & Wear
12/3830-543210 Printed on acid-free paper SPIN 10791483

Preface

Most of the introductory courses on linear algebra develop the basic theory of finite-dimensional vector spaces, and in so doing relate the notion of a linear mapping to that of a matrix. Generally speaking, such courses culminate in the diagonalisation of certain matrices and the application of this process to various situations. Such is the case, for example, in our previous SUMS volume *Basic Linear Algebra*. The present text is a continuation of that volume, and has the objective of introducing the reader to more advanced properties of vector spaces and linear mappings, and consequently of matrices. For readers who are not familiar with the contents of *Basic Linear Algebra* we provide an introductory chapter that consists of a compact summary of the prerequisites for the present volume.

In order to consolidate the student's understanding we have included a large number of illustrative and worked examples, as well as many exercises that are strategically placed throughout the text. Solutions to the exercises are also provided.

Many applications of linear algebra require careful, and at times rather tedious, calculations by hand. Very often these are subject to error, so the assistance of a computer is welcome. As far as computation in algebra is concerned, there are several packages available. Here we include, in the spirit of a tutorial, a chapter that gives a brief introduction to the use of MAPLE[1] in dealing with numerical and algebraic problems in linear algebra.

Finally, for the student who is keen on the history of mathematics, we include a chapter that contains brief biographies of those mathematicians mentioned in our two volumes on linear algebra. The biographies emphasise their contributions to linear algebra.

T.S.B., E.F.R.
St Andrews

---

[1] MAPLE™ is a registered trademark of Waterloo Maple Inc., 57 Erb Street West, Waterloo, Ontario, Canada, N2L 6C2. **www.maplesoft.com**

Contents

The story so far ....

Our title *Further Linear Algebra* suggests already that the reader will be familiar with the basics of this discipline. Since for each reader these can be different, depending on the content of the courses taken previously, we devote this introductory chapter to a compact summary of the prerequisites, thereby offering the reader a smooth entry to the material that follows. Readers already familiar with our previous SUMS volume *Basic Linear Algebra* may comfortably proceed immediately to Chapter 1.

In essence, linear algebra is the study of vector spaces. The notion of a vector space holds a ubiquitous place in mathematics and plays an important role, along with the attendant notion of a matrix, in many diverse applications.

If F is a given field then by a **vector space over** F, or an F-**vector space**, we mean an additive abelian group V together with an 'action' of F on V, described by $(\lambda, x) \mapsto \lambda x$, such that the following axioms hold:

(1) $(\forall \lambda \in F)(\forall x, y \in V) \quad \lambda(x + y) = \lambda x + \lambda y$;
(2) $(\forall \lambda, \mu \in F)(\forall x \in V) \quad (\lambda + \mu)x = \lambda x + \mu x$;
(3) $(\forall \lambda, \mu \in F)(\forall x \in V) \quad (\lambda \mu)x = \lambda(\mu x)$;
(4) $(\forall x \in V) \quad 1_F x = x$.

Additional properties that follow immediately from these axioms are:

(5) $(\forall \lambda \in F) \quad \lambda 0_V = 0_V$;
(6) $(\forall x \in V) \quad 0_F x = 0_V$;
(7) if $\lambda x = 0_V$ then either $\lambda = 0_F$ or $x = 0_V$;
(8) $(\forall \lambda \in F)(\forall x \in V) \quad (-\lambda)x = -(\lambda x) = \lambda(-x)$.

We often refer to the field F as the 'ground field' of the vector space V and the elements of F as 'scalars'. When F is the field $\mathbb{R}$ of real numbers we say that V is a **real** vector space; and when F is the field $\mathbb{C}$ of complex numbers we say that V is a **complex** vector space.

As with every algebraic structure, there are two important items to consider in relation to vector spaces, namely the *substructures* (i.e. the subsets that are also vector spaces) and the *morphisms* (i.e. the structure-preserving mappings between vector spaces).

As for the substructures, we say that a non-empty subset W of an F-vector space V is a **subspace** if it is closed under the operations of V, in the sense that

(1) if $x, y \in W$ then $x + y \in W$;
(2) if $x \in W$ and $\lambda \in F$ then $\lambda x \in W$.

These two conditions may be expressed as the single condition

$$(\forall \lambda, \mu \in F)(\forall x, y \in W) \quad \lambda x + \mu y \in W.$$

Every subspace of a vector space V contains 0_V, so $\{0_V\}$ is the smallest subspace. The set-theoretic intersection of any collection of subspaces of a vector space V is also a subspace of V. In particular, if S is any non-empty subset of V then there is a smallest subspace of V that contains S, namely the intersection of all the subspaces that contain S. This can be usefully identified as follows. We say that $v \in V$ is a **linear combination of elements of** S if there exist $x_1, \ldots, x_n \in S$ and $\lambda_1, \ldots, \lambda_n \in F$ such that

$$v = \sum_{i=1}^{n} \lambda_i x_i = \lambda_1 x_1 + \cdots + \lambda_n x_n.$$

Then the set of linear combinations of elements of S forms a subspace of V and is the smallest subspace of V to contain S. We call this the **subspace spanned by** S and denote it by Span S. We say that S is a **spanning set** of V if Span $S = V$.

A non-empty subset S of a vector space V is said to be **linearly independent** if the only way of writing 0_V as a linear combination of elements of S is the trivial way, in which all the scalars are 0_F. No linearly independent subset can contain 0_V. A subset that is not linearly independent is said to be **linearly dependent**. A subset S is linearly dependent if and only if at least one element of S can be expressed as a linear combination of the other elements of S.

A **basis** of a vector space V is a linearly independent subset that spans V. A non-empty subset S is a basis of V if and only if every element of V can be expressed in a unique way as a linear combination of elements of S.

If V is spanned by a finite subset $S = \{v_1, \ldots, v_n\}$ and if $I = \{w_1, \ldots, w_m\}$ is a linearly independent subset of V then necessarily $m \leqslant n$. From this there follows the important result that if V has a finite basis B then every basis of V is finite and has the same number of elements as B. A vector space V is said to be **finite-dimensional** if it has a finite basis. The number of elements in any basis of V is called the **dimension** of V and is written dim V. As a courtesy, we regard the empty set $\emptyset$ as a basis of the zero subspace $\{0_V\}$ which we can then say has dimension 0.

Every linearly independent subset of a finite-dimensional vector space V can be extended to form a basis of V. More precisely, if V is of dimension n and if

$$\{v_1, \ldots, v_m\}$$

is a linearly independent subset of V then there exist $v_{m+1}, \ldots, v_n \in V$ such that

$$\{v_1, \ldots, v_m, v_{m+1}, \ldots, v_n\}$$

is a basis of V.

If V is of finite dimension then so is every subspace W, and dim $W \leqslant$ dim V with equality if and only if $W = V$.

As for the morphisms, if V and W are vector spaces over the same field F then by a **linear mapping** from V to W we mean a mapping $f : V \to W$ such that

(1) $(\forall x, y \in V) \quad f(x + y) = f(x) + f(y)$;
(2) $(\forall \lambda \in F)(\forall x \in V) \quad f(\lambda x) = \lambda f(x)$.

These two conditions may be expressed as the single condition

$$(\forall \lambda, \mu \in F)(\forall x, y \in V) \quad f(\lambda x + \mu y) = \lambda f(x) + \mu f(y).$$

Additional properties that follow immediately from these axioms are:

(3) $f(0_V) = 0_W$;
(4) $(\forall x \in V) \quad f(-x) = -f(x)$.

If V and W are vector spaces over the same field F then the set Lin (V, W) of linear mappings from V to W can be made into a vector space over F in a natural way by defining $f + g$ and λf by the prescriptions

$$(\forall x \in V) \qquad (f + g)(x) = f(x) + g(x), \quad (\lambda f)(x) = \lambda f(x).$$

If $f : V \to W$ is linear and X is a subspace of V then the image of X under f, namely the set

$$f^{\to}(X) = \{f(x) \; ; \; x \in X\},$$

is a subspace of W; and if Y is a subspace of W then the pre-image of Y under f, namely the set

$$f^{\leftarrow}(Y) = \{x \in V \; ; f(x) \in Y\},$$

is a subspace of V. In particular, the subspace $f^{\to}(V) = \{f(x) \; ; \; x \in V\}$ of W is called the **image** of f and is written $\operatorname{Im} f$; and the subspace $f^{\leftarrow}\{0_W\}$ of V is called the **kernel** of f and is written $\operatorname{Ker} f$. A linear mapping $f : V \to W$ is injective if and only if $\operatorname{Ker} f = \{0_V\}$.

The main connection between $\operatorname{Im} f$ and $\operatorname{Ker} f$ is summarised in the **Dimension Theorem** which states that if V and W are finite-dimensional vector spaces over the same field and if $f : V \to W$ is linear then

$$\dim V = \dim \operatorname{Im} f + \dim \operatorname{Ker} f.$$

By the **rank** of a linear mapping f we mean $\dim \operatorname{Im} f$; and by the **nullity** of f we mean $\dim \operatorname{Ker} f$.

A bijective linear mapping is called an **isomorphism**. If V and W are vector spaces each of dimension n over a field F and if $f : V \to W$ is linear then the properties of being injective, surjective, bijective are equivalent, and f is an isomorphism if and only if f carries a basis to a basis.

If V is a vector space of dimension n over a field F then V is isomorphic to the vector space F^n over F that consists of n-tuples of elements of F. As a consequence, if V and W are vector spaces of the same dimension n over a field F then V and W are isomorphic.

Every linear mapping is completely and uniquely determined by its action on a basis. Thus, two linear mappings $f, g : V \to W$ are equal if and only if they agree on any basis of V.

Linear mappings from one finite-dimensional vector space to another can be represented by matrices. If F is a field and m, n are positive integers then by an $m \times n$ **matrix** over F we mean a rectangular array that consists of mn elements x_{ij} $(i = 1, \ldots, m; j = 1, \ldots, n)$ of F in a boxed display of m rows and n columns:

$$\begin{bmatrix} x_{11} & x_{12} & x_{13} & \cdots & x_{1n} \\ x_{21} & x_{22} & x_{23} & \cdots & x_{2n} \\ x_{31} & x_{32} & x_{33} & \cdots & x_{3n} \\ \vdots & \vdots & \vdots & \ddots & \vdots \\ x_{m1} & x_{m2} & x_{m3} & \cdots & x_{mn} \end{bmatrix}.$$

We often find it convenient to abbreviate the above display to simply $[x_{ij}]_{m\times n}$. The **transpose** of $A = [a_{ij}]_{m\times n}$ is the matrix $A^t = [a_{ji}]_{n\times m}$.

Matrices $A = [a_{ij}]_{m\times n}$ and $B = [b_{ij}]_{p\times q}$ are said to be **equal** if $m = p$, $n = q$ and $a_{ij} = b_{ij}$ for all i, j. With this definition of equality for matrices, the algebra of matrices may be summarised as follows. Given $m \times n$ matrices $A = [a_{ij}]_{m\times n}$ and $B = [b_{ij}]_{m\times n}$ over F, define $A + B = [a_{ij} + b_{ij}]_{m\times n}$; and for every $\lambda \in F$ define $\lambda A = [\lambda a_{ij}]_{m\times n}$. Then under these operations the set $\mathrm{Mat}_{m\times n} F$ of $m \times n$ matrices over F forms a vector space over F of dimension mn.

The connection with linear mappings is described as follows. First an **ordered basis** of a vector space V is a finite sequence $(v_i)_{1\leqslant i\leqslant n}$ of elements of V such that $\{v_1, \ldots, v_n\}$ is a basis of V. For convenience, we use the abbreviated notation $(v_i)_n$. Suppose now that V and W are vector spaces of dimensions m and n respectively over a field F. Let $(v_i)_m$, $(w_i)_n$ be given ordered bases of V, W. If $f : V \to W$ is linear then f is completely and uniquely determined by its action on the basis $(v_i)_m$. This action is described by expressing each $f(v_i)$ as a linear combination of the basis elements $w_1, \ldots, w_n$ of W:

$$f(v_1) = x_{11}w_1 + x_{12}w_2 + \cdots + x_{1n}w_n;$$
$$f(v_2) = x_{21}w_1 + x_{22}w_2 + \cdots + x_{2n}w_n;$$
$$\vdots$$
$$f(v_m) = x_{m1}w_1 + x_{m2}w_2 + \cdots + x_{mn}w_n.$$

The action of f on $(v_i)_m$ is therefore determined by the mn scalars x_{ij} appearing in the above equations. Put another way, the action of f is completely determined by a knowledge of the $m \times n$ matrix $X = [x_{ij}]$.

For technical reasons, the *transpose* of this matrix X is called the **matrix of f relative to the fixed ordered bases** $(v_i)_m, (w_i)_n$. When it is clear what these fixed ordered bases are, we denote the matrix in question by simply $\mathrm{Mat}\, f$.

If $f, g : V \to W$ are linear then relative to fixed ordered bases of each we have $\mathrm{Mat}\,(f + g) = \mathrm{Mat}\, f + \mathrm{Mat}\, g$, and $\mathrm{Mat}\,\lambda f = \lambda\, \mathrm{Mat}\, f$ for every scalar λ. Hence the assignment $f \mapsto \mathrm{Mat}\, f$ gives an isomorphism from the vector space $\mathrm{Lin}\,(V, W)$ to the vector space $\mathrm{Mat}_{m\times n} F$. In the particular case where $W = V$ the identity mapping id_V corresponds under this isomorphism to the **identity matrix** $I_n = [\delta_{ij}]_{n\times n}$ where

$$\delta_{ij} = \begin{cases} 1 & \text{if } i = j; \\ 0 & \text{otherwise.} \end{cases}$$

Consider now the following situation:

$$U; (u_i)_m \xrightarrow{\ f;A\ } V; (v_i)_n \xrightarrow{\ g;B\ } W; (w_i)_p$$

in which the notation $U; (u_i)_m$ for example denotes a vector space U with a fixed ordered basis $(u_i)_m$ and $f;A$ denotes a linear mapping f represented, relative to the ordered bases $(u_i)_m$, $(v_i)_n$, by the matrix A. The situation concerning the composite linear mapping $g \circ f : U \to W$ is

$$U; (u_i)_m \xrightarrow{\ g \circ f\ } W; (w_i)_p.$$

What is the matrix of this composite linear mapping? It is natural to expect that this will depend on the matrices A and B. In fact, with $A = [a_{ij}]_{n \times m}$ and $B = [b_{ij}]_{p \times n}$ (note the sizes!), we define the **product** BA to be the $p \times m$ matrix whose (i, j)-th element is

$$[BA]_{ij} = b_{i1}a_{1j} + b_{i2}a_{2j} + b_{i3}a_{3j} + \cdots + b_{in}a_{nj} = \sum_{k=1}^{n} b_{ik}a_{kj}.$$

Then we have that $\mathrm{Mat}\,(g \circ f) = BA = \mathrm{Mat}\,g \cdot \mathrm{Mat}\,f$. In other words, the matrix that represents $g \circ f$ is the product of the matrix that represents g with the matrix that represents f.

When the relevant sums and products are defined, we have the identities

$$A(BC) = (AB)C; \quad A(B + C) = AB + AC; \quad (B + C)A = BA + CA.$$

In particular, under the operations of addition and multiplication, the set $\mathrm{Mat}_{n \times n} F$ forms a ring with identity element I_n. Also, multiplication by scalars is such that, when the relevant products are defined,

$$\lambda(AB) = (\lambda A)B = A(\lambda B).$$

In particular, $\mathrm{Mat}_{n \times n} F$ becomes what is called an **algebra**.

Given a matrix $A = [a_{ij}]_{n \times m}$ over the field F, consider the subspace of $\mathrm{Mat}_{n \times 1} F$ that is spanned by the columns of A. The dimension of this subspace is called the **column rank** of A. Likewise, the dimension of the subspace of $\mathrm{Mat}_{1 \times m} F$ spanned by the rows of A is called the **row rank** of A. Somewhat surprisingly, the column rank and the row rank of a matrix over a field are the same. (This is not true for matrices over a ring, for example.) So we can talk simply of the **rank** of a matrix over a field.

An $n \times m$ matrix X is a **left inverse** of an $m \times n$ matrix M if $XM = I_n$; and a **right inverse** if $MX = I_m$. An $m \times n$ matrix M has a left inverse if and only if rank $M = n$, and has a right inverse if and only if rank $M = m$. A square matrix $M = [m_{ij}]_{n \times n}$ has a left inverse X if and only if it has a right inverse Y; moreover in this case $X = Y$ and so M has a unique **inverse** which we denote by M^{-1}. We say that a (necessarily square) matrix is **invertible** if it has an inverse.

An **elementary matrix** of size $n \times n$ is a matrix that is obtained from I_n by applying a single permutation to the rows or columns of I_n. If E is an elementary $n \times n$ matrix then for every $n \times n$ matrix A the matrices EA and AE have the same rank as A. A square matrix is invertible if and only if it is a product of elementary matrices.

If A and B are invertible then so is AB with $(AB)^{-1} = B^{-1}A^{-1}$. If A is invertible then so is its transpose, with $(A^t)^{-1} = (A^{-1})^t$. A matrix is **orthogonal** if A^{-1} exists and is A^t.

If V is a vector space over F of dimension n then a linear mapping $f : V \rightarrow V$ is an isomorphism if and only if it is represented, relative to some ordered basis, by an invertible matrix.

A very important question concerns what happens to the matrix of a linear mapping when we change reference from one ordered basis to another. Consider first the particular case of the identity mapping on an m-dimensional vector space V over F and the situation

$$V; (v_i)_m \xrightarrow{\;\; \mathrm{id}_V; A \;\;} V; (v'_i)_m$$

$$\uparrow \qquad\qquad\qquad \uparrow$$

$$\text{old basis} \qquad\qquad \text{new basis}$$

The matrix A that represents id_V is called the **transition matrix from the basis** $(v_i)_m$ **to the basis** $(v'_i)_m$. Clearly, transition matrices are invertible. If now W is an n-dimensional vector space over F and if a linear mapping $f : V \rightarrow W$ is represented relative to ordered bases $(v_i)_m, (w_i)_n$ by the $n \times m$ matrix A, what is the matrix that represents f relative to new ordered bases $(v'_i)_m, (w'_i)_n$? To solve this problem, which is the substance of the **Change of Basis Theorem**, we consider the situation

$$
\begin{array}{ccc}
V; (v_i)_m & \xrightarrow{\;\; f; A \;\;} & W; (w_i)_n \\[2pt]
\Big\uparrow {\scriptstyle \mathrm{id}_V; P} & & \Big\uparrow {\scriptstyle \mathrm{id}_W; Q} \\[2pt]
V; (v'_i)_m & \xrightarrow[\;\; f; X \;\;]{} & W; (w'_i)_n
\end{array}
$$

in which we have to determine the matrix X. Since, from the diagram, $f \circ \mathrm{id}_V = f = \mathrm{id}_W \circ f$ we deduce, via the isomorphism $\mathrm{Lin}(V, W) \simeq \mathrm{Mat}_{n \times m} F$, that $AP = QX$. Then, since the transition matrix Q is invertible, we have the solution $X = Q^{-1}AP$.

A direct consequence of this is that the rank of a linear mapping is the same as the rank of any matrix that represents it. In fact, let V, W be of dimensions m, n and let $f : V \dashrightarrow W$ be of rank p. Then the Dimension Theorem gives $\dim \mathrm{Ker}\, f = m - p$. Let $\{v_1, \ldots, v_{m-p}\}$ be a basis of $\mathrm{Ker}\, f$ and extend this to a basis

$$B = \{u_1, \ldots, u_p, v_1, \ldots, v_{m-p}\}$$

of V. Then $\{f(u_1), \ldots, f(u_p)\}$ is linearly independent and so is a basis for $\mathrm{Im}\, f$. Now extend this to a basis

$$C = \{f(u_1), \ldots, f(u_p), w_1, \ldots, w_{n-p}\}$$

of W. Then the matrix of f relative to the ordered bases B, C is

$$\begin{bmatrix} I_p & 0 \\ 0 & 0 \end{bmatrix}.$$

Suppose now that A is an $n \times m$ matrix that represents f relative to fixed ordered bases B_V, B_W. If Q and P are the transition matrices from the bases B, C to the bases B_V, B_W then we have

$$Q^{-1}AP = \begin{bmatrix} I_p & 0 \\ 0 & 0 \end{bmatrix},$$

the matrix on the right being of rank p. But since transition matrices are invertible they are products of elementary matrices. Consequently the rank of A is the same as the rank of $Q^{-1}AP$, namely $p = \operatorname{rank} f$.

If A, B are $n \times n$ matrices over F then B is said to be **similar** to A if there is an invertible matrix P such that $B = P^{-1}AP$. The relation of being similar is an equivalence relation on the set $\operatorname{Mat}_{n \times n} F$. Concerning similarity an important problem, from both the theoretical and practical points of view, is the determination of particularly simple representatives (or **canonical forms**) in each similarity class. The starting point of this investigation involves the notion of a **diagonal matrix**, this being a square matrix $A = [a_{ij}]_{n \times n}$ for which $a_{ij} = 0$ whenever $i \neq j$. The problem of deciding when a given square matrix is similar to a diagonal matrix is equivalent to that of deciding when a linear mapping can be represented by a diagonal matrix. To tackle this problem, we need the following machinery which involves the theory of **determinants**. We shall not summarise this theory here, but refer the reader to Chapter 8 of our *Basic Linear Algebra* for a succinct treatment.

If A is an $n \times n$ matrix then an **eigenvalue** of A is a scalar λ for which there exists a *non-zero* $n \times 1$ matrix $\mathbf{x}$ such that $A\mathbf{x} = \lambda\mathbf{x}$. Such a (column) matrix $\mathbf{x}$ is called an **eigenvector** associated with λ. A scalar λ is an eigenvalue of A if and only if $\det(A - \lambda I_n) = 0$. Here $\det(A - XI_n)$ is a polynomial of degree n in X, called the **characteristic polynomial**, and $\det(A - XI_n) = 0$ is the **characteristic equation**. The **algebraic multiplicity** of an eigenvalue λ is the greatest integer k such that $(X - \lambda)^k$ is a factor of the characteristic polynomial.

The notion of characteristic polynomial can be defined for a linear mapping. Given a vector space V of dimension n over F and a linear mapping $f : V \to V$, let A be the matrix of f relative to some fixed ordered basis of V. Then the matrix of f relative to any other ordered basis is of the form $P^{-1}AP$ where P is the transition matrix from the new basis to the old basis. Now the characteristic polynomial of $P^{-1}AP$ is

$$\det(P^{-1}AP - XI_n) = \det\left[P^{-1}(A - XI_n)P\right]$$

$$= \det P^{-1} \cdot \det(A - XI_n) \cdot \det P$$

$$= \det(A - XI_n),$$

i.e. we have $c_{P^{-1}AP}(X) = c_A(X)$. It follows that the characteristic polynomial is independent of the choice of basis, so we can define the characteristic polynomial of f to be the characteristic polynomial of any matrix that represents f.

If λ is an eigenvalue of the $n \times n$ matrix A then the set

$$E_\lambda = \{\mathbf{x} \in \mathrm{Mat}_{n \times 1} F \; ; \; A\mathbf{x} = \lambda\mathbf{x}\}$$

i.e. the set of eigenvectors associated with the eigenvalue λ together with the zero column $\mathbf{0}$, is a subspace of the vector space $\mathrm{Mat}_{n \times 1} F$. This subspace is called the **eigenspace** associated with the eigenvalue λ. The dimension of the eigenspace E_λ is called the **geometric multiplicity** of the eigenvalue λ.

If $f : V \to W$ is a linear mapping then a scalar λ is said to be an **eigenvalue** of f if there is a *non-zero* $x \in V$ such that $f(x) = \lambda x$, such an element x being called an **eigenvector** associated with λ. The connection with matrices is as follows. Given an $n \times n$ matrix A, consider the linear mapping

$$f_A : \mathrm{Mat}_{n \times 1} F \to \mathrm{Mat}_{n \times 1} F$$

given by $f_A(\mathbf{x}) = A\mathbf{x}$. Relative to the natural ordered basis of $\mathrm{Mat}_{n \times 1} F$ (given by the columns of I_n), we have $\mathrm{Mat}\, f_A = A$. Clearly, the matrix A and the linear mapping f_A have the same eigenvalues.

Eigenvectors that correspond to distinct eigenvalues are linearly independent.

A square matrix is said to be **diagonalisable** if it is similar to a diagonal matrix. An $n \times n$ matrix is diagonalisable if and only if it admits n linearly independent eigenvectors. A linear mapping $f : V \to V$ is said to be **diagonalisable** if there is an ordered basis $(v_i)_n$ of V with repect to which the matrix of f is a diagonal matrix; equivalently, if and only if V has a basis consisting of eigenvectors of f.

A square matrix A [resp. a linear mapping f] is diagonalisable if and only if, for every eigenvalue λ, the geometric multiplicity of λ coincides with its algebraic multiplicity.

If a given matrix A is similar to a diagonal matrix D then there is an invertible matrix P such that $P^{-1}AP = D$ where the diagonal entries of D are the eigenvalues of A. The practical problem of determining such a matrix P is dealt with as follows. First we observe that the equation $P^{-1}AP = D$ can be written $AP = PD$. Let the columns of P be $\mathbf{p}_1, \ldots, \mathbf{p}_n$ and let

$$D = \begin{bmatrix} \lambda_1 & & & \\ & \lambda_2 & & \\ & & \ddots & \\ & & & \lambda_n \end{bmatrix}$$

where $\lambda_1, \ldots, \lambda_n$ are the eigenvalues of A. Comparing the i-th columns of each side of the equation $AP = PD$, we obtain

$$(i = 1, \ldots, n) \qquad A\mathbf{p}_i = \lambda_i \mathbf{p}_i.$$

In other words, the i-th column of P is an eigenvector of A corresponding to the eigenvalue λ_i. So, when a given $n \times n$ matrix A is diagonalisable, to determine an invertible matrix P that will reduce A to diagonal form (more precisely, such that

$P^{-1}AP = \text{diag}\{\lambda_1, \ldots, \lambda_n\}$) we simply determine an eigenvector corresponding to each of the n eigenvalues then paste these eigenvectors together as the columns of the matrix P.

Consider again the vector space $\text{Mat}_{n \times n} F$ which is of dimension n^2. Every set of $n^2 + 1$ elements of $\text{Mat}_{n \times n} F$ must be linearly dependent. In particular, given any $A \in \text{Mat}_{n \times n} F$, the $n^2 + 1$ powers

$$A^0 = I_n, \ A, \ A^2, \ A^3, \ \ldots, A^{n^2}$$

are linearly dependent and so there is a non-zero polynomial

$$p(X) = a_0 + a_1 X + a_2 X^2 + \cdots + a_{n^2} X^{n^2}$$

with coefficients in F such that $p(A) = 0$. The same is of course true for any $f \in \text{Lin}(V, V)$ where V is of dimension n, for we have $\text{Lin}(V, V) \simeq \text{Mat}_{n \times n} F$.

But a significantly better result holds: there is in fact a polynomial $p(X)$ which is *of degree at most n* and such that $p(A) = 0$. This is the celebrated **Cayley–Hamilton Theorem**, the polynomial in question being the characteristic polynomial $c_A(X) = \det(A - XI_n)$ of A.

It follows from this that there is a unique monic polynomial $m_A(X)$ of least degree such that $m_A(A) = 0$. This is called the **minimum polynomial** of A. Both $m_A(X)$ and $c_A(X)$ have the same zeros, so $m_A(X)$ divides $c_A(X)$.

In a similar way, we can define the notion of the minimum polynomial of a linear mapping, namely as the minimum polynomial of any matrix that represents the mapping.

Inner Product Spaces

In some aspects of vector spaces the ground field F of scalars can be arbitrary. In this chapter, however, we shall restrict F to be $\mathbb{R}$ or $\mathbb{C}$, for the results that we obtain depend heavily on the properties of these fields.

Definition

Let V be a vector space over $\mathbb{C}$. By an **inner product** on V we shall mean a mapping $f : V \times V \to \mathbb{C}$, described by $(x, y) \mapsto \langle x \mid y \rangle$, such that for all $x, x', y \in V$ and all $\alpha \in \mathbb{C}$, the following identities hold:

(1) $\langle x + x' \mid y \rangle = \langle x \mid y \rangle + \langle x' \mid y \rangle$;

(2) $\langle \alpha x \mid y \rangle = \alpha \langle x \mid y \rangle$;

(3) $\overline{\langle x \mid y \rangle} = \langle y \mid x \rangle$ so that in particular $\langle x \mid x \rangle \in \mathbb{R}$;

(4) $\langle x \mid x \rangle \geqslant 0$, with equality if and only if $x = 0_V$.

By a **complex inner product space** we mean a vector space V over $\mathbb{C}$ together with an inner product on V. By a **real inner product space** we mean a vector space V over $\mathbb{R}$ together with an inner product on V (this being defined as in the above but with the bar denoting complex conjugate omitted). By an **inner product space** we shall mean either a complex inner product space or a real inner product space.

There are other useful identities that follow immediately from (1) to (4) above, namely:

(5) $\langle x \mid y + y' \rangle = \langle x \mid y \rangle + \langle x \mid y' \rangle$.

In fact, by (1) and (3) we have

$$\langle x \mid y + y' \rangle = \overline{\langle y + y' \mid x \rangle} = \overline{\langle y \mid x \rangle} + \overline{\langle y' \mid x \rangle} = \langle x \mid y \rangle + \langle x \mid y' \rangle.$$

(6) $\langle x \mid \alpha y \rangle = \overline{\alpha} \langle x \mid y \rangle$.

This follows from (3) and (4) since

$$\langle x \mid \alpha y \rangle = \overline{\langle \alpha y \mid x \rangle} = \overline{\alpha \langle y \mid x \rangle} = \overline{\alpha} \, \overline{\langle y \mid x \rangle} = \overline{\alpha} \langle x \mid y \rangle.$$

(7) $\langle x \mid 0_V \rangle = 0 = \langle 0_V \mid x \rangle$.

This is immediate from (1), (2), (5) on taking $x' = -x$, $y' = -y$, and $\alpha = -1$.

Example 1.1

On the vector space $\mathbb{R}^n$ of n-tuples of real numbers let

$$\langle (x_1, \ldots, x_n) \mid (y_1, \ldots, y_n) \rangle = \sum_{i=1}^{n} x_i y_i.$$

Then it is readily verified that this defines an inner product on $\mathbb{R}^n$, called the **standard inner product** on $\mathbb{R}^n$.

In the cases where $n = 2, 3$ this inner product is often called the **dot product** or **scalar product**. This terminology is popular when dealing with the geometric applications of vectors. Indeed, several of the results that we shall establish will generalise familiar results in euclidean geometry of two and three dimensions.

Example 1.2

On the vector space $\mathbb{C}^n$ of n-tuples of complex numbers let

$$\langle (z_1, \ldots, z_n) \mid (w_1, \ldots, w_n) \rangle = \sum_{i=1}^{n} z_i \overline{w_i}.$$

Then it is readily verified that this defines an inner product on $\mathbb{C}^n$, called the **standard inner product** on $\mathbb{C}^n$.

Example 1.3

Let $a, b \in \mathbb{R}$ with $a < b$ and let V be the real vector space of continuous functions $f : [a, b] \to \mathbb{R}$. Define a mapping from $V \times V$ to $\mathbb{R}$ by

$$(f, g) \mapsto \langle f \mid g \rangle = \int_a^b fg.$$

Then, by elementary properties of integrals, this defines an inner product on V.

EXERCISES

1.1 Let $\mathbb{R}_n[X]$ be the real vector space of polynomials at degree most n. Prove that

$$\langle p \mid q \rangle = \int_0^1 pq$$

defines an inner product on $\mathbb{R}_n[X]$.

1.2 For a square matrix $A = [a_{ij}]_{n \times n}$ define its **trace** by $\operatorname{tr} A = \sum_{i=1}^{n} a_{ii}$. Prove that the vector space $\operatorname{Mat}_{n \times n} \mathbb{R}$ can be made into a real inner product space by defining

$$\langle A \mid B \rangle = \operatorname{tr} B^t A.$$

Likewise, prove that $\operatorname{Mat}_{n \times n} \mathbb{C}$ can be made into a complex inner product space by defining

$$\langle A \mid B \rangle = \operatorname{tr} B^\star A$$

where $B^\star = \overline{B^t}$ is the complex conjugate of the transpose of B.

Definition

Let V be an inner product space. For every $x \in V$ we define the **norm** of x to be the non-negative real number

$$\|x\| = \sqrt{\langle x \mid x \rangle}.$$

For $x, y \in V$ we define the **distance between** x **and** y to be

$$d(x, y) = \|x - y\|.$$

Example 1.4

In the real inner product space $\mathbb{R}^2$ under the standard inner product, if $x = (x_1, x_2)$ then $\|x\|^2 = x_1^2 + x_2^2$, so $\|x\|$ is the distance from x to the origin. Likewise, if $y = (y_1, y_2)$ then we have

$$\|x - y\|^2 = (x_1 - y_1)^2 + (x_2 - y_2)^2,$$

which is simply the theorem of Pythagoras.

Recall from (4) above that we have

$$\|x\| = 0 \iff x = 0_V.$$

Theorem 1.1

Let V be an inner product space. Then, for all $x, y \in V$ and every scalar λ,

(1) $\|\lambda x\| = |\lambda| \, \|x\|$;

(2) [Cauchy–Schwarz inequality] $|\langle x \mid y \rangle| \leqslant \|x\| \, \|y\|$;

(3) [Triangle inequality] $\|x + y\| \leqslant \|x\| + \|y\|$.

Proof

(1) $\|\lambda x\|^2 = \langle \lambda x \mid \lambda x \rangle = \lambda \overline{\lambda} \langle x \mid x \rangle = |\lambda|^2 \|x\|^2$.

(2) The result is trivial if $x = 0_V$. Suppose then that $x \neq 0_V$, so that $\|x\| \neq 0$. Let $z = y - \dfrac{\langle y \mid x \rangle}{\|x\|^2} x$. Then

$$\langle z \mid x \rangle = \langle y \mid x \rangle - \frac{\langle y \mid x \rangle}{\|x\|^2} \langle x \mid x \rangle = 0$$

and therefore

$$
\begin{aligned}
0 \leqslant \|z\|^2 = \langle z \mid z \rangle &= \langle z \mid y \rangle \\
&= \langle y \mid y \rangle - \frac{\langle y \mid x \rangle}{\|x\|^2} \langle x \mid y \rangle \\
&= \langle y \mid y \rangle - \frac{|\langle x \mid y \rangle|^2}{\|x\|^2}
\end{aligned}
$$

from which (2) follows.

(3) We have

$$\|x + y\|^2 = \langle x + y \,|\, x + y \rangle$$
$$= \langle x \,|\, x \rangle + \langle x \,|\, y \rangle + \langle y \,|\, x \rangle + \langle y \,|\, y \rangle$$
$$= \|x\|^2 + \langle x \,|\, y \rangle + \overline{\langle x \,|\, y \rangle} + \|y\|^2$$
$$= \|x\|^2 + 2 \,\mathrm{Re}\langle x \,|\, y \rangle + \|y\|^2$$
$$\leqslant \|x\|^2 + 2|\langle x \,|\, y \rangle| + \|y\|^2$$
$$\leqslant \|x\|^2 + 2 \,\|x\| \,\|y\| + \|y\|^2$$
$$= (\,\|x\| + \|y\|\,)^2,$$

from which the triangle equality follows immediately. $\square$

Example 1.5

In the inner product space $\mathbb{C}^n$ under the standard inner product, the Cauchy–Schwarz inequality becomes

$$\left| \sum_{k=1}^{n} \alpha_k \overline{\beta_k} \right| \leqslant \sqrt{\sum_{k=1}^{n} |\alpha_k|^2} \sqrt{\sum_{k=1}^{n} |\beta_k|^2}.$$

Example 1.6

Let V be the set of sequences $(a_i)_{i \geqslant 1}$ of real numbers that are **square-summable** in the sense that $\sum_{i \geqslant 1} a_i^2$ exists. Define an addition and a multiplication by real scalars in the component-wise manner, namely let

$$(a_i)_{i \geqslant 1} + (b_i)_{i \geqslant 1} = (a_i + b_i)_{i \geqslant 1}, \quad \lambda(a_i)_{i \geqslant 1} = (\lambda a_i)_{i \geqslant 1}.$$

Then it is readily verified that under these operations V becomes a real vector space.

Now by the Cauchy–Schwarz inequality applied to the inner product space $\mathbb{R}^n$ with the standard inner product, we have the inequality

$$\left| \sum_{i=1}^{n} a_i b_i \right| \leqslant \sqrt{\sum_{i=1}^{n} a_i^2 \sum_{i=1}^{n} b_i^2}.$$

If now $(p_n)_{n \geqslant 1}, (q_n)_{n \geqslant 1} \in V$ are given by $p_n = \sum_{i=1}^{n} a_i b_i$ and $q_n = \sqrt{\sum_{i=1}^{n} a_i^2 \sum_{i=1}^{n} b_i^2}$ then

we have $|p_n| \leqslant q_n$ where $(q_n)_{n \geqslant 1}$ converges to $\sqrt{\sum_{i \geqslant 1} a_i^2 \sum_{i \geqslant 1} b_i^2}$. It follows that $(p_n)_{n \geqslant 1}$ is absolutely convergent and hence is convergent. Thus $\sum_{i \geqslant 1} a_i b_i$ exists and we can define $\langle (a_i)_{i \geqslant 1} \,|\, (b_i)_{i \geqslant 1} \rangle = \sum_{i \geqslant 1} a_i b_i$.

In this way, V becomes a real inner product space that is often called ℓ_2-**space** or **Hilbert space**.

EXERCISES

1.3 If V is a real inner product space and $x, y \in V$, prove that

$$\|x + y\|^2 = \|x\|^2 + \|y\|^2 + 2\langle x \mid y \rangle.$$

Interpret this result in $\mathbb{R}^2$.

1.4 If W is a complex inner product space and $x, y \in W$, prove that

$$\|x + y\|^2 - i\|ix + y\|^2 = \|x\|^2 + \|y\|^2 - i(\|x\|^2 + \|y\|^2) + 2\langle x \mid y \rangle.$$

1.5 If V is an inner product space establish the **parallelogram identity**

$$(\forall x, y \in V) \quad \|x + y\|^2 + \|x - y\|^2 = 2\|x\|^2 + 2\|y\|^2.$$

Interpret this result in $\mathbb{R}^2$.

1.6 If V is a real inner product space and $x, y \in V$ are such that $\|x\| = \|y\|$, show that $\langle x + y \mid x - y \rangle = 0$.
Interpret this result in $\mathbb{R}^2$.

1.7 Interpret the Cauchy–Schwarz inequality in the inner product space of Example 1.3.

1.8 Interpret the Cauchy–Schwarz inequality in the complex inner product space $\text{Mat}_{n \times n} \mathbb{C}$ of Exercise 1.2. If, in this inner product space, E_{pq} denotes the matrix whose (p, q)-th entry is 1 and all other entries are 0, determine $\langle E_{pq} \mid E_{rs} \rangle$.

1.9 Show that equality holds in the Cauchy–Schwarz inequality if and only if $\{x, y\}$ is linearly dependent.

Definition

If V is an inner product space then $x, y \in V$ are said to be **orthogonal** if $\langle x \mid y \rangle = 0$. A non-empty subset S of V is said to be an **orthogonal subset** if every pair of distinct elements of S is orthogonal. An **orthonormal subset** of V is an orthogonal subset S such that $\|x\| = 1$ for every $x \in S$.

Example 1.7

Relative to the standard inner products, the standard bases of $\mathbb{R}^n$ and of $\mathbb{C}^n$ form orthonormal subsets.

Example 1.8

The matrices $E_{p,q}$ of Exercise 1.8 form an orthonormal subset of the complex inner product space $\text{Mat}_{n \times n} \mathbb{C}$.

Example 1.9

In $\mathbb{R}^2$ the elements $x = (x_1, x_2)$ and $y = (y_1, y_2)$ are orthogonal if and only if $x_1 y_1 + x_2 y_2 = 0$. Geometrically, this is equivalent to saying that the lines joining x and y to the origin are mutually perpendicular.

EXERCISES

1.10 In the inner product space $\mathbb{R}_2[X]$ of Exercise 1.1, determine for which $a \in \mathbb{R}$ the subset $\{X, X^2 - a\}$ is orthogonal.

1.11 Let V be the inner product space of real continuous functions on the interval $[-\pi, \pi]$ with inner product

$$\langle f \mid g \rangle = \int_{-\pi}^{\pi} fg.$$

Prove that the subset

$$S = \{x \mapsto 1, \ x \mapsto \sin kx, \ x \mapsto \cos kx \ ; \ k = 1, 2, 3, \dots \}$$

is orthogonal.

1.12 In the inner product space $\text{Mat}_{2 \times 2} \mathbb{C}$ with $\langle A \mid B \rangle = \text{tr } B^\star A$, is the subset

$$S = \left\{ \begin{bmatrix} i & 0 \\ 0 & 0 \end{bmatrix}, \begin{bmatrix} 0 & i \\ 0 & 0 \end{bmatrix}, \begin{bmatrix} 0 & 0 \\ i & 0 \end{bmatrix}, \begin{bmatrix} 0 & 0 \\ 0 & i \end{bmatrix} \right\}$$

orthonormal?

Clearly, an orthonormal subset of V can always be constructed from an orthogonal subset S consisting of non-zero elements of V by the process of **normalising** each element of S, i.e. by replacing $x \in S$ by $x^\star = \dfrac{x}{\|x\|}$.

An important property of orthonormal sets is the following.

Theorem 1.2

In an inner product space orthogonal subsets of non-zero elements are linearly independent.

Proof

Let S be an orthogonal subset of $V \setminus \{0_V\}$ and let $x_1, \dots, x_n \in S$. Suppose that $\sum_{i=1}^{n} \lambda_i x_i = 0_V$. Then we have

$$\lambda_i \|x_i\|^2 = \lambda_i \langle x_i \mid x_i \rangle = \sum_{k=1}^{n} \lambda_k \langle x_k \mid x_i \rangle = \left\langle \sum_{k=1}^{n} \lambda_k x_k \mid x_i \right\rangle = \langle 0_V \mid x_i \rangle = 0,$$

whence each $\lambda_i = 0$ since $x_i \neq 0_V$. Hence S is linearly independent. $\square$

We now describe the subspace that is spanned by an orthonormal subset.

Theorem 1.3

Let $\{e_1, \ldots, e_n\}$ be an orthonormal subset of the inner product space V. Then

$$[\text{Bessel's inequality}] \qquad (\forall x \in V) \quad \sum_{k=1}^{n} |\langle x \,|\, e_k \rangle|^2 \leqslant \|x\|^2.$$

Moreover, if W is the subspace spanned by $\{e_1, \ldots, e_n\}$ then the following statements are equivalent:

(1) $x \in W$;

(2) $\sum_{k=1}^{n} |\langle x \,|\, e_k \rangle|^2 = \|x\|^2$;

(3) $x = \sum_{k=1}^{n} \langle x \,|\, e_k \rangle e_k$;

(4) $(\forall y \in V) \quad \langle x \,|\, y \rangle = \sum_{k=1}^{n} \langle x \,|\, e_k \rangle \langle e_k \,|\, y \rangle$.

Proof

Let $z = x - \sum_{k=1}^{n} \langle x \,|\, e_k \rangle e_k$. Then a simple calculation gives

$$\langle z \,|\, z \rangle = \langle x \,|\, x \rangle - \sum_{k=1}^{n} \langle x \,|\, e_k \rangle \overline{\langle x \,|\, e_k \rangle} = \|x\|^2 - \sum_{k=1}^{n} |\langle x \,|\, e_k \rangle|^2.$$

$(2) \Rightarrow (3)$ is now immediate from this since if (2) holds then $z = 0_V$.

$(3) \Rightarrow (4)$: If $x = \sum_{k=1}^{n} \langle x \,|\, e_k \rangle e_k$ then for all $y \in V$ we have

$$\langle x \,|\, y \rangle = \left\langle \sum_{k=1}^{n} \langle x \,|\, e_k \rangle e_k \,\Big|\, y \right\rangle = \sum_{k=1}^{n} \langle x \,|\, e_k \rangle \langle e_k \,|\, y \rangle.$$

$(4) \Rightarrow (2)$ follows by taking $y = x$ in (4).

$(3) \Rightarrow (1)$ is clear.

$(1) \Rightarrow (3)$: If $x = \sum_{k=1}^{n} \lambda_k e_k$ then for $j = 1, \ldots, n$ we have

$$\lambda_j = \sum_{k=1}^{n} \lambda_k \langle e_k \,|\, e_j \rangle = \left\langle \sum_{k=1}^{n} \lambda_k e_k \,\Big|\, e_j \right\rangle = \langle x \,|\, e_j \rangle$$

whence (3) holds. $\quad \square$

Definition

By an **orthonormal basis** of an inner product space we mean an orthonormal subset that is a basis.

Example 1.10

The standard bases of $\mathbb{R}^n$ and of $\mathbb{C}^n$ are orthonormal bases.

Example 1.11

In $\text{Mat}_{n \times n} \mathbb{C}$ with $\langle A \,|\, B \rangle = \text{tr } B^{\star}A$ an orthonormal basis is $\{E_{p,q} \; ; \; p, q = 1, \ldots, n\}$ where $E_{p,q}$ has a 1 in the (p, q)-th position and 0 elsewhere.

We shall now show that every finite-dimensional inner product space contains an orthonormal basis. In so doing, we give a practical method of constructing such a basis.

Theorem 1.4

[Gram–Schmidt orthonormalisation process]   *Let V be an inner product space and for every non-zero $x \in V$ let $x^{\star} = x/\|x\|$. If $\{x_1, \ldots, x_k\}$ is a linearly independent subset of V, define recursively*

$$y_1 = x_1^{\star};$$

$$y_2 = \left(x_2 - \langle x_2 \,|\, y_1 \rangle y_1\right)^{\star};$$

$$y_3 = \left(x_3 - \langle x_3 \,|\, y_2 \rangle y_2 - \langle x_3 \,|\, y_1 \rangle y_1\right)^{\star};$$

$$\vdots$$

$$y_k = \left(x_k - \sum_{i=1}^{k-1} \langle x_k \,|\, y_i \rangle y_i\right)^{\star}.$$

Then $\{y_1, \ldots, y_k\}$ is orthonormal and spans the same subspace as $\{x_1, \ldots, x_k\}$.

Proof

Since $x_1 \neq 0_V$ it is clear that $y_1 \neq 0_V$. Then $x_2 - \langle x_2 \,|\, y_1 \rangle y_1$ is a non-trivial linear combination of x_2, x_1 and hence $y_2 \neq 0_V$. In general, y_i is a non-trivial linear combination of $x_i, \ldots, x_1$ and consequently $y_i \neq 0_V$. It is clear from the definition of y_i that x_i is a linear combination of $y_1, \ldots, y_i$. It follows therefore that $\{x_1, \ldots, x_k\}$ and $\{y_1, \ldots, y_k\}$ span the same subspace of V.

It now suffices to prove that $\{y_1, \ldots, y_k\}$ is an orthogonal subset; and this we do inductively. For $k = 1$ the result is trivial. Suppose then that $\{y_1, \ldots, y_{r-1}\}$ is orthogonal where $r > 2$. Then, writing

$$\left\| x_r - \sum_{i=1}^{r-1} \langle x_r \,|\, y_i \rangle y_i \right\| = \alpha_i,$$

we see from the definition of y_r that

$$\alpha_r y_r = x_r - \sum_{i=1}^{r-1} \langle x_r \,|\, y_i \rangle y_i$$

and so, for $j < r$,

$$\alpha_r \langle y_r | y_j \rangle = \langle x_r | y_j \rangle - \sum_{i=1}^{r-1} \langle x_r | y_i \rangle \langle y_i | y_j \rangle$$

$$= \langle x_r | y_j \rangle - \langle x_r | y_j \rangle$$

$$= 0.$$

Since $\alpha_r \neq 0$ we deduce that $\langle y_r | y_j \rangle = 0$ for $j < r$. Consequently, $\{y_1, \dots, y_r\}$ is orthogonal. $\square$

Corollary

If V is a finite-dimensional inner product space then V has an orthonormal basis.

Proof

Apply the Gram–Schmidt process to a basis of V. $\square$

Example 1.12

Consider the basis $\{x_1, x_2, x_3\}$ of $\mathbb{R}^3$ where

$$x_1 = (0, 1, 1), \quad x_2 = (1, 0, 1), \quad x_3 = (1, 1, 0).$$

In order to apply the Gram–Schmidt process to this basis using the standard inner product, we first let

$$y_1 = x_1 / \|x_1\| = \tfrac{1}{\sqrt{2}}(0, 1, 1).$$

Then

$$x_2 - \langle x_2 | y_1 \rangle y_1 = (1, 0, 1) - \tfrac{1}{\sqrt{2}} \langle (1, 0, 1) | (0, 1, 1) \rangle \tfrac{1}{\sqrt{2}}(0, 1, 1)$$

$$= (1, 0, 1) - \tfrac{1}{2}(0, 1, 1)$$

$$= \tfrac{1}{2}(2, -1, 1)$$

so, normalising this, we take $y_2 = \tfrac{1}{\sqrt{6}}(2, -1, 1)$.

[Note that, by Theorem 1.1, we have in general

$$(\lambda x)^* = \frac{\lambda x}{\|\lambda x\|} = \frac{\lambda x}{|\lambda| \, \|x\|} = \begin{cases} x^* & \text{if } \lambda > 0; \\ -x^* & \text{if } \lambda < 0. \end{cases}$$

This helps avoid unnecessary arithmetic in the calculations.]

Similarly,

$$x_3 - \langle x_3 | y_2 \rangle y_2 - \langle x_3 | y_1 \rangle y_1 = \tfrac{2}{3}(1, 1, -1)$$

and so, normalising this, we take $y_3 = \tfrac{1}{\sqrt{3}}(1, 1, -1)$.

Thus an orthonormal basis constructed from the given basis is

$$\{ \tfrac{1}{\sqrt{2}}(0, 1, 1), \tfrac{1}{\sqrt{6}}(2, -1, 1), \tfrac{1}{\sqrt{3}}(1, 1, -1) \}.$$

Theorem 1.5

If $\{e_1, \ldots, e_n\}$ is an orthonormal basis of an inner product space V then

(1) $(\forall x \in V) \qquad x = \sum_{k=1}^{n} \langle x \mid e_k \rangle e_k;$

(2) $(\forall x \in V) \qquad \|x\|^2 = \sum_{k=1}^{n} |\langle x \mid e_k \rangle|^2;$

(3) $(\forall x, y \in V) \quad \langle x \mid y \rangle = \sum_{k=1}^{n} \langle x \mid e_k \rangle \langle e_k \mid y \rangle.$

Proof

This is immediate from Theorem 1.3. $\square$

Definition

The identity (1) in Theorem 1.5 is often called the **Fourier expansion of** x **relative to the orthonormal basis** $\{e_1, \ldots, e_n\}$, the scalars $\langle x \mid e_k \rangle$ being called the **Fourier coefficients** of x. The identity (3) of Theorem 1.5 is known as **Parseval's identity**.

Example 1.13

The Fourier coefficients of $(1, 1, 1)$ relative to the orthonormal basis $\{y_1, y_2, y_3\}$ constructed in the previous example are

$$\langle (1, 1, 1) \mid \tfrac{1}{\sqrt{2}}(0, 1, 1) \rangle = \sqrt{2};$$

$$\langle (1, 1, 1) \mid \tfrac{1}{\sqrt{6}}(2, -1, 1) \rangle = \tfrac{\sqrt{2}}{\sqrt{3}};$$

$$\langle (1, 1, 1) \mid \tfrac{1}{\sqrt{3}}(1, 1, -1) \rangle = \tfrac{1}{\sqrt{3}}.$$

Just as every linearly independent subset can be extended to a basis, so can every orthonormal subset be extended to an orthonormal basis.

Theorem 1.6

Let V be an inner product space of dimension n. If $\{x_1, \ldots, x_k\}$ is an orthonormal subset of V then there exist $x_{k+1}, \ldots, x_n \in V$ such that $\{x_1, \ldots, x_n\}$ is an orthonormal basis of V.

Proof

Let W be the subspace spanned by $\{x_1, \ldots, x_k\}$. By Theorem 1.2, this set is a basis for W and so can be extended to a basis $\{x_1, \ldots, x_k, x_{k+1}, \ldots, x_n\}$ of V. Applying the Gram–Schmidt process to this basis, we obtain an orthonormal basis for V. The result now follows on noting that the first k elements of this orthonormal basis are precisely $x_1, \ldots, x_k$; for, by the formulae in Theorem 1.4 and the orthogonality of $x_1, \ldots, x_k$, we see that $y_i = x_i^\star = x_i$ for each $i \leqslant k$. $\square$

EXERCISES

1.13 Use the Gram–Schmidt process to construct in $\mathbb{R}^3$ an orthonormal basis from the basis

$$\{(1,1,1,),(0,1,1),(0,0,1)\}.$$

1.14 Determine an orthonormal basis for the subspace of $\mathbb{R}^4$ spanned by

$$\{(1,1,0,1),(1,-2,0,0),(1,0,-1,2)\}.$$

1.15 Determine an orthonormal basis for the subspace of $\mathbb{C}^3$ spanned by

$$\{(1,i,0),(1,2,1-i)\}.$$

1.16 Consider the real inner product space of real continuous functions on the interval $[0,1]$ under the inner product

$$\langle f \mid g \rangle = \int_0^1 fg.$$

Find an orthonormal basis for the subspace spanned by $\{f_1, f_2\}$ where $f_1(x) = 1$ and $f_2(x) = x$.

1.17 In the inner product space $\mathbb{R}[X]$ with inner product

$$\langle p \mid q \rangle = \int_{-1}^1 pq$$

determine an orthonormal basis for the subspace spanned by

(1) $\{X, X^2\}$;
(2) $\{1, 2X - 1, 12X^2\}$.

1.18 Let V be a real inner product space. Given $x, y \in V$ define the **angle** between x and y to be the real number $\vartheta \in [0, \pi]$ such that

$$\cos \vartheta = \frac{\langle x \mid y \rangle}{\|x\| \, \|y\|}.$$

Prove that, in the real inner product space $\mathbb{R}^2$, Parseval's identity gives

$$\cos(\vartheta_1 - \vartheta_2) = \cos \vartheta_1 \cos \vartheta_2 + \sin \vartheta_1 \sin \vartheta_2.$$

1.19 Referring to Exercise 1.11, consider the orthogonal subset

$$S_n = \{x \mapsto 1, \ x \mapsto \sin kx, \ x \mapsto \cos kx \ ; \ k = 1, \dots, n\}.$$

If T_n is the orthonormal subset obtained by normalising every element of S_n, interpret Bessel's inequality relative to T_n.

Determine, relative to T_n, the Fourier expansion of

(1) $x \mapsto x^2$;
(2) $x \mapsto \sin x + \cos x$.

1.20 Show that

$$B = \{(\tfrac{1}{\sqrt{5}}, 0, \tfrac{2}{\sqrt{5}}), (-\tfrac{2}{\sqrt{5}}, 0, \tfrac{1}{\sqrt{5}}), (0, 1, 0)\}$$

is an orthonormal basis of $\mathbb{R}^3$. Express the vector $(2, -3, 1)$ as a linear combination of elements of B.

1.21 Let V be a real inner product space and let $f : V \to V$ be linear. Prove that if $B = \{e_1, \ldots, e_n\}$ is an ordered orthonormal basis of V and if M is the matrix of f relative to B then

$$[M]_{ij} = \langle f(e_j) \mid e_i \rangle.$$

1.22 Let V be a real inner product space and let $A = \{a_1, \ldots, a_r\}$ be a subset of V. Define G_A to be the $r \times r$ matrix whose (i, j)-th element is

$$[G_A]_{ij} = \langle a_i \mid a_j \rangle.$$

Prove that $a_1, \ldots, a_r$ are linearly dependent if and only if $\det G_A = 0$.

An isomorphism from one finite-dimensional vector space to another carries bases to bases. We now investigate the corresponding situation for inner product spaces.

Definition

If V and W are inner product spaces over the same field then $f : V \to W$ is an **inner product isomorphism** if it is a vector space isomorphism that preserves inner products, in the sense that

$$(\forall x, y \in V) \qquad \langle f(x) \mid f(y) \rangle = \langle x \mid y \rangle.$$

Theorem 1.7

Let V and W be finite-dimensional inner product spaces over the same field. Let $\{e_1, \ldots, e_n\}$ be an orthonormal basis of V. Then $f : V \to W$ is an inner product isomorphism if and only if $\{f(e_1), \ldots, f(e_n)\}$ is an orthonormal basis of W.

Proof

$\Rightarrow$: If $f : V \to W$ is an inner product isomorphism then clearly $\{f(e_1), \ldots, f(e_n)\}$ is a basis of W. It is also orthonormal since

$$\langle f(e_i) \mid f(e_j) \rangle = \langle e_i \mid e_j \rangle = \begin{cases} 1 & \text{if } i = j; \\ 0 & \text{if } i \neq j. \end{cases}$$

$\Leftarrow$: Suppose now that $\{f(e_1), \ldots, f(e_n)\}$ is an orthonormal basis of W. Then f carries a basis of V to a basis of W and so f is a vector space isomorphism. Now for all $x \in V$ we have, using the Fourier expansion of x relative to the orthonormal basis $\{e_1, \ldots, e_n\}$,

$$\begin{aligned}
\langle f(x) \,|\, f(e_j) \rangle &= \left\langle f\left(\sum_{i=1}^{n} \langle x \,|\, e_i \rangle e_i \right) \,\middle|\, f(e_j) \right\rangle \\
&= \left\langle \sum_{i=1}^{n} \langle x \,|\, e_i \rangle f(e_i) \,\middle|\, f(e_j) \right\rangle \\
&= \sum_{i=1}^{n} \langle x \,|\, e_i \rangle \langle f(e_i) \,|\, f(e_j) \rangle \\
&= \langle x \,|\, e_j \rangle
\end{aligned}$$

and similarly $\langle f(e_j) \,|\, f(x) \rangle = \langle e_j \,|\, x \rangle$. It now follows by Parseval's identity applied to both V and W that

$$\begin{aligned}
\langle f(x) \,|\, f(y) \rangle &= \sum_{j=1}^{n} \langle f(x) \,|\, f(e_j) \rangle \langle f(e_j) \,|\, f(y) \rangle \\
&= \sum_{j=1}^{n} \langle x \,|\, e_j \rangle \langle e_j \,|\, y \rangle \\
&= \langle x \,|\, y \rangle
\end{aligned}$$

and consequently f is an inner product isomorphism. $\square$

EXERCISES

1.23 Let V and W be finite-dimensional inner product spaces over the same field with $\dim V = \dim W$. If $f : V \to W$ is linear, prove that the following statements are equivalent:

(1) f is an inner product isomorphism;

(2) f preserves norms, i.e. $(\forall x, \in V)$ $\|f(x)\| = \|x\|$;

(3) f preserves distances, i.e. $(\forall x, y \in V)$ $d(f(x), f(y)) = d(x, y)$.

1.24 Let V be the infinite-dimensional vector space of real continuous functions on the interval $[0, 1]$. Let V_1 consist of V with the inner product

$$\langle f \,|\, g \rangle = \int_0^1 f(x) g(x) x^2 \, dx,$$

and let V_2 consist of V with the inner product

$$\langle f \,|\, g \rangle = \int_0^1 f(x) g(x) \, dx.$$

Prove that the mapping $\varphi : V_1 \to V_2$ given by $f \mapsto \varphi(f)$ where

$$(\forall x \in V) \qquad [\varphi(f)](x) = x f(x)$$

is linear and preserves inner products, but is not an inner product isomorphism.

2

Direct Sums of Subspaces

If A and B are non-empty subsets of a vector space V over a field F then the subspace spanned by $A \cup B$, i.e. the smallest subspace of V that contains both A and B, is the set of linear combinations of elements of $A \cup B$. In other words, it is the set of elements of the form

$$\sum_{i=1}^{m} \lambda_i a_i + \sum_{j=1}^{n} \mu_j b_j$$

where $a_i \in A$, $b_j \in B$, and $\lambda_i, \mu_j \in F$. In the particular case where A and B are subspaces of V we have $\sum_{i=1}^{m} \lambda_i a_i \in A$ and $\sum_{j=1}^{n} \mu_j b_j \in B$, so this set can be described as

$$\{a + b \; ; \; a \in A, \, b \in B\}.$$

We call this the **sum** of the subspaces A, B and denote it by $A + B$. This operation of addition on subspaces is clearly commutative. More generally, if $A_1, \ldots, A_n$ are subspaces of V then we define their **sum** to be the subspace that is spanned by $\bigcup_{i=1}^{n} A_i$.

We denote this subspace by $A_1 + \cdots + A_n$ or $\sum_{i=1}^{n} A_i$, noting that the operation of addition on subspaces is also associative. Clearly, we have

$$\sum_{i=1}^{n} A_i = \{a_1 + \cdots + a_n \; ; \; a_i \in A_i\}.$$

Example 2.1

Let X, Y, D be the subspaces of $\mathbb{R}^2$ given by

$$X = \{(x,0) \; ; \; x \in \mathbb{R}\}, \quad Y = \{(0,y) \; ; \; y \in \mathbb{R}\}, \quad D = \{(x,x) \; ; \; x \in \mathbb{R}\}.$$

Since every $(x,y) \in \mathbb{R}^2$ can be written in each of the three ways

$$(x,0) + (0,y), \quad (x-y,0) + (y,y), \quad (0,y-x) + (x,x),$$

we see that $\mathbb{R}^2 = X + Y = X + D = Y + D$.

Definition

A sum $\sum_{i=1}^{n} A_i$ of subspaces $A_1, \ldots, A_n$ is said to be **direct** if every $x \in \sum_{i=1}^{n} A_i$ can be written in a unique way as $x = a_1 + \cdots + a_n$ with $a_i \in A_i$ for each i.

When the sum $\sum_{i=1}^{n} A_i$ is direct we shall write it as $\bigoplus_{i=1}^{n} A_i$ or $A_1 \oplus \cdots \oplus A_n$ and call this the **direct sum** of the subspaces $A_1, \ldots, A_n$.

Example 2.2

In Example 2.1 we have $\mathbb{R}^2 = X \oplus Y = X \oplus D = Y \oplus D$.

Example 2.3

In $\mathbb{R}^3$ consider the subspaces

$$A = \{(x, y, z) \; ; \; x + y + z = 0\}, \quad B = \{(x, x, z) \; ; \; x, z \in \mathbb{R}\}.$$

Then $\mathbb{R}^3 = A + B$ since, for example, we can write each (x, y, z) as

$$\left(\tfrac{1}{2}(x - y), -\tfrac{1}{2}(x - y), 0 \right) + \left(\tfrac{1}{2}(x + y), \tfrac{1}{2}(x + y), z \right)$$

which is of the form $a + b$ where $a \in A$ and $b \in B$. The sum $A + B$ is not direct, however, since such an expression is not unique. To see this, observe that (x, y, z) can also be expressed as

$$\left(\tfrac{1}{2}(x - y + 1), -\tfrac{1}{2}(x - y - 1), -1 \right) + \left(\tfrac{1}{2}(x + y - 1), \tfrac{1}{2}(x + y - 1), z + 1 \right)$$

which is of the form $a' + b'$ with $a' \in A$, $b' \in B$ and $a' \neq a$, $b' \neq b$.

EXERCISES

2.1 Let V be the subspace of $\mathbb{R}^3$ given by

$$V = \{(x, x, 0) \; ; \; x \in \mathbb{R}\}.$$

Find distinct subspaces U_1, U_2 of $\mathbb{R}^3$ such that $\mathbb{R}^3 = V \oplus U_1 = V \oplus U_2$.

2.2 Let $t_1, t_2, t_3, t_4 : \mathbb{R}^3 \to \mathbb{R}^3$ be the linear mappings given by

$$t_1(x, y, z) = (x + y, y + z, z + x);$$

$$t_2(x, y, z) = (x - y, y - z, 0);$$

$$t_3(x, y, z) = (-y, x, z);$$

$$t_4(x, y, z) = (x, y, y).$$

Prove that, for $i = 1, 2, 3, 4$,

$$\mathbb{R}^3 = \operatorname{Im} t_i \oplus \operatorname{Ker} t_i.$$

A useful criterion for a sum to be direct is provided by the following result.

Theorem 2.1

If $A_1, \ldots, A_n$ are subspaces of a vector space V then the following statements are equivalent:

(1) *the sum $\sum_{i=1}^{n} A_i$ is direct;*

(2) *if $\sum_{i=1}^{n} a_i = 0$ with $a_i \in A_i$ then each $a_i = 0$;*

(3) *$A_i \cap \sum_{j \neq i} A_j = \{0_V\}$ for every i.*

Proof

$(1) \Rightarrow (2)$: By the definition of direct sum, if (1) holds then 0_V can be written in precisely one way as a sum $\sum_{i=1}^{n} a_i$ with $a_i \in A_i$ for each i. Condition (2) follows immediately from this.

$(2) \Rightarrow (3)$: Suppose that (2) holds and let $x \in A_i \cap \sum_{j \neq i} A_j$, say $x = a_i = \sum_{j \neq i} a_j$. We can write this as $a_i - \sum_{j \neq i} a_j = 0$ whence, by (2), each $a_i = 0$ and therefore $x = 0_V$.

$(3) \Rightarrow (1)$: Suppose now that (3) holds and that $\sum_{i=1}^{n} a_i = \sum_{i=1}^{n} b_i$ where $a_i, b_i \in A_i$ for each i. Then we have

$$a_i - b_i = \sum_{j \neq i} (b_j - a_j)$$

where the left hand side belongs to A_i and the right hand side belongs to $\sum_{j \neq i} A_j$. By (3), we deduce that $a_i - b_i = 0$ and, since this holds for each i, (1) follows. $\square$

Corollary

$V = A_1 \oplus A_2$ if and only if $V = A_1 + A_2$ and $A_1 \cap A_2 = \{0_V\}$. $\square$

Example 2.4

A mapping $f : \mathbb{R} \to \mathbb{R}$ is said to be **even** if $f(-x) = f(x)$ for every $x \in \mathbb{R}$; and **odd** if $f(-x) = -f(x)$ for every $x \in \mathbb{R}$. The set A of even functions is a subspace of the vector space $V = \text{Map}(\mathbb{R}, \mathbb{R})$, as is the set B of odd functions. Moreover, $V = A \oplus B$. To see this, given any $f : \mathbb{R} \to \mathbb{R}$ let $f^+ : \mathbb{R} \to \mathbb{R}$ and $f^- : \mathbb{R} \to \mathbb{R}$ be given by

$$f^+(x) = \tfrac{1}{2}[f(x) + f(-x)], \quad f^-(x) = \tfrac{1}{2}[f(x) - f(-x)].$$

Then f^+ is even and f^- is odd. Since clearly $f = f^+ + f^-$ we have $V = A + B$. Now $A \cap B$ consists of only the zero mapping and so, by the above corollary, we have that $V = A \oplus B$.

Example 2.5

Let V be the vector space $\text{Mat}_{n\times n}\,\mathbb{R}$. If A, B are respectively the subspaces of V that consist of the symmetric and skew-symmetric matrices then $V = A \oplus B$. For, every $M \in V$ can be written uniquely in the form $X + Y$ where X is symmetric and Y is skew-symmetric; as is readily seen, $X = \frac{1}{2}(M + M')$ and $Y = \frac{1}{2}(M - M')$.

EXERCISES

2.3 If V is a finite-dimensional vector space and if A is a non-zero subspace of V prove that there exists a subspace B of V such that $V = A \oplus B$.

2.4 If $X = \{(x,0)\ ;\ x \in \mathbb{R}\}$ find infinitely many subspaces Z_i such that $\mathbb{R}^2 = X \oplus Z_i$.

2.5 If V is a finite-dimensional vector space and A, B are subspaces of V such that $V = A + B$ and $\dim V = \dim A + \dim B$, prove that $V = A \oplus B$.

2.6 Let V be a finite-dimensional vector space and let $f : V \to V$ be a linear mapping. Prove that $V = \text{Im} f \oplus \text{Ker} f$ if and only if $\text{Im} f = \text{Im} f^2$.

2.7 If V is a finite-dimensional vector space and $f : V \to V$ is a linear mapping, prove that there is a positive integer p such that $\text{Im} f^p = \text{Im} f^{p+1}$ and deduce that

$$(\forall k \geqslant 1) \qquad \text{Im} f^p = \text{Im} f^{p+k}, \quad \text{Ker} f^p = \text{Ker} f^{p+k}.$$

Show also that $V = \text{Im} f^p \oplus \text{Ker} f^p$.

A significant property of direct sums is the following, which says informally that bases of the summands can be pasted together to obtain a basis for their sum.

Theorem 2.2

Let $V_1, \ldots, V_n$ be non-zero subspaces of a finite-dimensional vector space V such that $V = \bigoplus\limits_{i=1}^{n} V_i$. If B_i is a basis of V_i then $\bigcup\limits_{i=1}^{n} B_i$ is a basis of V.

Proof

For each i, let the subspace V_i have dimension d_i and basis $B_i = \{b_{i,1}, \ldots, b_{i,d_i}\}$. Since $V = \bigoplus\limits_{i=1}^{n} V_i$ we have $V_i \cap \sum\limits_{j\neq i} V_j = \{0_V\}$ and therefore $V_i \cap V_j = \{0_V\}$ for $i \neq j$. Consequently $B_i \cap B_j = \emptyset$ for $i \neq j$. Now a typical element of the subspace spanned by $\bigcup\limits_{i=1}^{n} B_i$ is of the form

$$(1) \qquad \sum_{j=1}^{d_1} \lambda_{1j} b_{1,j} + \cdots + \sum_{j=1}^{d_n} \lambda_{nj} b_{n,j}$$

i.e. of the form

$$(2) \qquad x_1 + \cdots + x_n \quad \text{where} \quad x_i = \sum_{j=1}^{d_i} \lambda_{ij} b_{i,j}.$$

Since $V = \sum_{i=1}^{n} V_i$ and since B_i is a basis of V_i it is clear that every $x \in V$ can be expressed in the form (1), and so V is spanned by $\bigcup_{i=1}^{n} B_i$. If now in (2) we have $x_1 + \cdots + x_n = 0_V$ then by Theorem 2.1 we deduce that each $x_i = 0_V$ and consequently that each $\lambda_{ij} = 0$. Thus $\bigcup_{i=1}^{n} B_i$ is a basis of V. $\square$

Corollary

$$\dim \bigoplus_{i=1}^{n} V_i = \sum_{i=1}^{n} \dim V_i. \quad \square$$

EXERCISES

2.8 Let V be a real vector space of dimension 4 and let $B = \{b_1, b_2, b_3, b_4\}$ be a basis of V. With each $x \in V$ expressed as $x = \sum_{i=1}^{4} x_i b_i$, let

$$V_1 = \{x \in V ; \; x_3 = x_2, \; x_4 = x_1\};$$
$$V_2 = \{x \in V ; \; x_3 = -x_2, \; x_4 = -x_1\}.$$

Show that

(1) V_1 and V_2 are subspaces of V;

(2) $B_1 = \{b_1 + b_4, b_2 + b_3\}$ is a basis of V_1, and $B_2 = \{b_2 - b_3, b_1 - b_4\}$ is a basis of V_2;

(3) $V = V_1 \oplus V_2$;

(4) the transition matrix from the basis B to the basis $B_1 \cup B_2$ is

$$P = \begin{bmatrix} \frac{1}{2} & 0 & 0 & \frac{1}{2} \\ 0 & \frac{1}{2} & \frac{1}{2} & 0 \\ 0 & \frac{1}{2} & -\frac{1}{2} & 0 \\ \frac{1}{2} & 0 & 0 & -\frac{1}{2} \end{bmatrix} ;$$

(5) $P^{-1} = 2P$.

A real 4×4 matrix M is said to be **centro-symmetric** if $m_{ij} = m_{5-i, 5-j}$ for all i, j. If M is centro-symmetric, prove that M is similar to a matrix of the form

$$\begin{bmatrix} \alpha & \beta & 0 & 0 \\ \gamma & \delta & 0 & 0 \\ 0 & 0 & \varepsilon & \zeta \\ 0 & 0 & \eta & \vartheta \end{bmatrix}.$$

2.9 Let V be a vector space of dimension n over $\mathbb{R}$. If $f : V \to V$ is linear and such that $f^2 = \mathrm{id}_V$ prove that

$$V = \mathrm{Im}(\mathrm{id}_V + f) \oplus \mathrm{Im}(\mathrm{id}_V - f).$$

Deduce that an $n \times n$ matrix A over $\mathbb{R}$ is such that $A^2 = I_n$ if and only if A is similar to a matrix of the form

$$\begin{bmatrix} I_p & 0 \\ 0 & -I_{n-p} \end{bmatrix}.$$

Our purpose now is to determine precisely when a vector space is the direct sum of finitely many non-zero subspaces. As we shall see, this is closely related with the following types of linear mapping.

Definition

Let A and B be subspaces of a vector space V such that $V = A \oplus B$. Then every $x \in V$ can be expressed uniquely in the form $x = a + b$ where $a \in A$ and $b \in B$. By the **projection on A parallel to B** we shall mean the linear mapping $p_A : V \to V$ given by $p_A(x) = a$.

Example 2.6

We know that $\mathbb{R}^2 = X \oplus D$ where $X = \{(x, 0) \; ; \; x \in \mathbb{R}\}$ and $D = \{(x, x) \; ; \; x \in \mathbb{R}\}$. The projection on X parallel to D is given by $p(x, y) = (x - y, 0)$. Thus the image of the point (x, y) is the point of intersection with X of the line through (x, y) parallel to the line D. The terminology used is thus suggested by the geometry.

EXERCISE

2.10 In $\mathbb{R}^3$ let $A = \mathrm{Span}\{(1, 0, 1), (-1, 1, 2)\}$. Determine the projection of $(1, 2, 1)$ onto A parallel to $\mathrm{Span}\{(0, 1, 0)\}$.

Definition

A linear mapping $f : V \to V$ is said to be a **projection** if there are subspaces A, B of V such that $V = A \oplus B$ and f is the projection on A parallel to B. A linear mapping $f : V \to V$ is said to be **idempotent** if $f^2 = f$.

These notions are related as follows.

Theorem 2.3

If $V = A \oplus B$ and if p is the projection on A parallel to B then

(1) $A = \mathrm{Im}\, p = \{x \in V \; ; \; x = p(x)\}$;
(2) $B = \mathrm{Ker}\, p$;
(3) *p is idempotent.*

Proof

(1) It is clear that $A = \operatorname{Im} p \supseteq \{x \in V ; \ x = p(x)\}$. If now $a \in A$ then its unique representation as a sum of an element of A and an element of B is clearly $a = a + 0_V$. Consequently $p(a) = a$ and the containment becomes equality.

(2) Let $x \in V$ have the unique representation $x = a + b$ where $a \in A$ and $b \in B$. Then since $p(x) = a$ we have

$$p(x) = 0_V \iff a = 0_V \iff x = b \in B.$$

In other words, $\operatorname{Ker} p = B$.

(3) For every $x \in V$ we have $p(x) \in A$ and so, by (1), $p(x) = p[p(x)]$. Thus $p^2 = p$. $\square$

That projections and idempotents are essentially the same is the substance of the following result.

Theorem 2.4

A linear mapping $f : V \to V$ is a projection if and only if it is idempotent, in which case $V = \operatorname{Im} f \oplus \operatorname{Ker} f$ and f is the projection on $\operatorname{Im} f$ parallel to $\operatorname{Ker} f$.

Proof

Suppose that f is a projection. Then there exist subspaces A and B with $V = A \oplus B$, and f is the projection on A parallel to B. By Theorem 2.3(3), f is idempotent.

Conversely, suppose that $f : V \to V$ is idempotent. If $x \in \operatorname{Im} f \cap \operatorname{Ker} f$ then we have $x = f(y)$ for some y, and $f(x) = 0_V$. Consequently, $x = f(y) = f[f(y)] = f(x) = 0_V$ and hence $\operatorname{Im} f \cap \operatorname{Ker} f = \{0_V\}$. Now for every $x \in V$ we observe that

$$f[x - f(x)] = f(x) - f^2(x) = f(x) - f(x) = 0_V$$

and so $x - f(x) \in \operatorname{Ker} f$. The identity $x = f(x) + x - f(x)$ now shows that we also have $V = \operatorname{Im} f + \operatorname{Ker} f$. It follows by the Corollary of Theorem 2.1 that $V = \operatorname{Im} f \oplus \operatorname{Ker} f$. Suppose now that $x = a + b$ where $a \in \operatorname{Im} f$ and $b \in \operatorname{Ker} f$. Then $a = f(y)$ for some y, and $f(b) = 0_V$. Consequently,

$$f(x) = f(a + b) = f(a) + 0_V = f[f(y)] = f(y) = a.$$

Thus we see that f is the projection on $\operatorname{Im} f$ parallel to $\operatorname{Ker} f$. $\square$

Corollary

If $f : V \to V$ is a projection then so is $\operatorname{id}_V - f$, and $\operatorname{Im} f = \operatorname{Ker}(\operatorname{id}_V - f)$.

Proof

Since $f^2 = f$ we have $(\operatorname{id}_V - f)^2 = \operatorname{id}_V - f - f + f^2 = \operatorname{id}_V - f$. Moreover, by Theorem 2.3, we have

$$x \in \operatorname{Im} f \iff x = f(x) \iff (\operatorname{id}_V - f)(x) = 0_V$$

and so $\operatorname{Im} f = \operatorname{Ker}(\operatorname{id}_V - f)$. $\square$

EXERCISES

2.11 If f is the projection on A parallel to B, prove that $\mathrm{id}_V - f$ is the projection on B parallel to A.

2.12 Let V be a vector space over $\mathbb{R}$. If $p_1, p_2 : V \to V$ are projections prove that the following are equivalent:

(1) $p_1 + p_2$ is a projection;

(2) $p_1 \circ p_2 = p_2 \circ p_1 = 0$.

When $p_1 + p_2$ is a projection, prove that $\mathrm{Im}(p_1 + p_2) = \mathrm{Im}\, p_1 \oplus \mathrm{Im}\, p_2$ and $\mathrm{Ker}(p_1 + p_2) = \mathrm{Ker}\, p_1 \cap \mathrm{Ker}\, p_2$.

2.13 If V is a finite-dimensional vector space and if $p_1, p_2 : V \to V$ are projections, prove that the following statements are equivalent:

(1) $\mathrm{Im}\, p_1 = \mathrm{Im}\, p_2$;

(2) $p_1 \circ p_2 = p_2$ and $p_2 \circ p_1 = p_1$.

2.14 Let V be a finite-dimensional vector space and let $p_1, \ldots, p_k : V \to V$ be projections such that $\mathrm{Im}\, p_1 = \mathrm{Im}\, p_2 = \cdots = \mathrm{Im}\, p_k$. Let $\lambda_1, \ldots, \lambda_k \in F$ be such that $\sum_{i=1}^{k} \lambda_i = 1$. Prove that $p = \lambda_1 p_1 + \cdots + \lambda_k p_k$ is a projection with $\mathrm{Im}\, p = \mathrm{Im}\, p_i$.

2.15 Let V be a vector space over $\mathbb{R}$ and let $f : V \to V$ be idempotent. Prove that $\mathrm{id}_V + f$ is invertible and determine its inverse.

We now show how the decomposition of a vector space into a direct sum of finitely many non-zero subspaces may be expressed in terms of projections.

Theorem 2.5

If V is a vector space then there are non-zero subspaces $V_1, \ldots, V_n$ of V such that $V = \bigoplus_{i=1}^{n} V_i$ if and only if there are non-zero linear mappings $p_1, \ldots, p_n : V \to V$ such that

(1) $\sum_{i=1}^{n} p_i = \mathrm{id}_V$;

(2) $(i \neq j)\ p_i \circ p_j = 0$.

Moreover, each p_i is necessarily a projection and $V_i = \mathrm{Im}\, p_i$.

Proof

Suppose first that $V = \bigoplus_{i=1}^{n} V_i$. Then for $i = 1, \ldots, n$ we have $V = V_i \oplus \sum_{j \neq i} V_j$. Let p_i be the projection on V_i parallel to $\sum_{j \neq i} V_j$. Using the notation $p_i^{\rightarrow}(X) = \{p_i(x)\ ;\ x \in X\}$

for every subspace X of V, we have that for $j \neq i$,

$$p_i[p_j(x)] \in p_i^{\rightarrow}(\operatorname{Im} p_j) = p_i^{\rightarrow}(V_j) \quad \text{by Theorem 2.3}$$

$$\subseteq p_i^{\rightarrow}\left(\sum_{k \neq i} V_k\right)$$

$$= p_i^{\rightarrow}(\operatorname{Ker} p_i) \quad \text{by Theorem 2.4}$$

$$= \{0_V\}$$

and so $p_i \circ p_j = 0$. Also, since every $x \in V$ can be written uniquely in the form $x = \sum_{i=1}^{n} x_i$ where $x_i \in V_i$ for each i, we have

$$x = \sum_{i=1}^{n} x_i = \sum_{i=1}^{n} p_i(x) = \left(\sum_{i=1}^{n} p_i\right)(x)$$

whence $\sum_{i=1}^{n} p_i = \operatorname{id}_V$.

Conversely, suppose that $p_1, \ldots, p_n$ satisfy (1) and (2). Then we note that

$$p_i = p_i \circ \operatorname{id}_V = p_i \circ \sum_{j=1}^{n} p_j = \sum_{j=1}^{n} (p_i \circ p_j) = p_i \circ p_i$$

and so each p_i is idempotent and therefore, by Theorem 2.4, is a projection.

Now for every $x \in V$ we have

$$x = \operatorname{id}_V(x) = \left(\sum_{i=1}^{n} p_i\right)(x) = \sum_{i=1}^{n} p_i(x) \in \sum_{i=1}^{n} \operatorname{Im} p_i$$

which shows that $V = \sum_{i=1}^{n} \operatorname{Im} p_i$.

If now $x \in \operatorname{Im} p_i \cap \sum_{j \neq i} \operatorname{Im} p_j$ then, by Theorem 2.3, we have $x = p_i(x)$ and $x = \sum_{j \neq i} x_j$ where $p_j(x_j) = x_j$ for every $j \neq i$. Consequently,

$$x = p_i(x) = p_i\left(\sum_{j \neq i} x_j\right) = p_i\left(\sum_{j \neq i} p_j(x_j)\right) = \sum_{j \neq i} p_i[p_j(x)] = 0_V$$

and it follows that $V = \bigoplus_{i=1}^{n} \operatorname{Im} p_i$. $\square$

Example 2.7

Consider the direct sum decomposition $\mathbb{R}^2 = X \oplus Y$ where X is the 'x-axis' and Y is the 'y-axis'. The projection on X parallel to Y is given by $p_X(x, y) = (x, 0)$, and the projection on Y parallel to X is given by $p_Y(x, y) = (0, y)$. We have

$$p_X + p_Y = \operatorname{id}, \qquad p_X \circ p_Y = 0 = p_Y \circ p_X.$$

We now pass to the consideration of direct sums of subspaces in inner product spaces. Although we shall take a closer look at this later, there is a particular direct sum of subspaces that we can deal with immediately. For this purpose, consider the following notion.

Definition

Let V be an inner product space. For every non-empty subset E of V we define the **orthogonal complement** of E to be the set $E^\perp$ of elements of V that are orthogonal to every element of E; in symbols,

$$E^\perp = \{y \in V ; \; (\forall x \in E) \; \langle x \,|\, y \rangle = 0\}.$$

It is readily verified that $E^\perp$ is a subspace of V. Moreover, we have $V^\perp = \{0_V\}$ and $\{0_V\}^\perp = V$.

The above terminology is suggested by the following result.

Theorem 2.6

Let V be an inner product space and let W be a finite-dimensional subspace of V. Then

$$V = W \oplus W^\perp.$$

Proof

Since W is of finite dimension there exists, by the Corollary of Theorem 1.4, an orthonormal basis of W, say $\{e_1, \ldots, e_n\}$. Given $x \in V$, let $x' = \sum_{i=1}^{n} \langle x \,|\, e_i \rangle e_i \in W$ and consider the element $x'' = x - x'$. For $j = 1, \ldots, n$ we have

$$\langle x'' \,|\, e_j \rangle = \langle x \,|\, e_j \rangle - \langle x' \,|\, e_j \rangle = \langle x \,|\, e_j \rangle - \sum_{i=1}^{n} \langle x \,|\, e_i \rangle \langle e_i \,|\, e_j \rangle$$

$$= \langle x \,|\, e_j \rangle - \langle x \,|\, e_j \rangle$$

$$= 0.$$

It follows that $x'' \in W^\perp$ and hence that $x = x' + x'' \in W + W^\perp$. Consequently we have $V = W + W^\perp$.

Now if $x \in W \cap W^\perp$ we have $\langle x \,|\, x \rangle = 0$ whence $x = 0_V$. Thus $W \cap W^\perp = \{0_V\}$ and we conclude that $V = W \oplus W^\perp$. $\square$

The principal properties of orthogonal complements are contained in the following two results.

Theorem 2.7

If V is a finite-dimensional inner product space and if W is a subspace of V then

$$\dim W^\perp = \dim V - \dim W.$$

Moreover, $(W^\perp)^\perp = W$.

Proof

It is immediate from Theorem 2.6 that

$$\dim V = \dim W + \dim W^{\perp},$$

from which the first statement follows. As for the second, it is clear from the definition of orthogonal complement that we have $W \subseteq (W^{\perp})^{\perp}$. Also, by what we have just proved,

$$\dim (W^{\perp})^{\perp} = \dim V - \dim W^{\perp}.$$

Hence $\dim (W^{\perp})^{\perp} = \dim W$ and the result follows. $\square$

Theorem 2.8

If V is a finite-dimensional inner product space and if A, B are subspaces of V then

(1) $A \subseteq B \implies B^{\perp} \subseteq A^{\perp}$;
(2) $(A \cap B)^{\perp} = A^{\perp} + B^{\perp}$;
(3) $(A + B)^{\perp} = A^{\perp} \cap B^{\perp}$.

Proof

(1) If $A \subseteq B$ then clearly every element of V that is orthogonal to B is also orthogonal to A, so $B^{\perp} \subseteq A^{\perp}$.

(2) Since $A, B \subseteq A + B$ we have, by (1),

$$(A + B)^{\perp} \subseteq A^{\perp} \cap B^{\perp}.$$

Similarly, since $A \cap B \subseteq A, B$ we have $A^{\perp}, B^{\perp} \subseteq (A \cap B)^{\perp}$ and so

$$A^{\perp} + B^{\perp} \subseteq (A \cap B)^{\perp}.$$

Together with Theorem 2.7, these observations give

$$A \cap B = (A \cap B)^{\perp\perp} \subseteq (A^{\perp} + B^{\perp})^{\perp} \subseteq A^{\perp\perp} \cap B^{\perp\perp} = A \cap B$$

from which we deduce that

$$A \cap B = (A^{\perp} + B^{\perp})^{\perp}.$$

Consequently $(A \cap B)^{\perp} = (A^{\perp} + B^{\perp})^{\perp\perp} = A^{\perp} + B^{\perp}$.

(3) This follows from (2) on replacing A, B by $A^{\perp}, B^{\perp}$. $\square$

Example 2.8

Consider the subspace W of $\mathbb{R}^3$ given by $W = \text{Span}\{(0, 1, 1), (1, 0, 1)\}$. In order to determine $W^{\perp}$ we can proceed as follows. First extend $\{(0, 1, 1), (1, 0, 1)\}$ to the basis $\{(0, 1, 1), (1, 0, 1), (1, 1, 0)\}$ of $\mathbb{R}^3$. Now apply the Gram–Schmidt process to obtain (as in Example 1.12) the orthonormal basis

$$\left\{ \tfrac{1}{\sqrt{2}}(0, 1, 1), \tfrac{1}{\sqrt{6}}(2, -1, 1), \tfrac{1}{\sqrt{3}}(1, 1, -1) \right\}.$$

Recall from Theorem 1.4 that $W = \text{Span}\left\{\frac{1}{\sqrt{2}}(0,1,1), \frac{1}{\sqrt{6}}(2,-1,1)\right\}$. It follows that $W^\perp = \text{Span}\{(1,1,-1)\}$.

EXERCISE

2.16 Determine the orthogonal complement in $\mathbb{R}^3$ of the subspace

$$\text{Span}\{(1,0,1),(1,2,-2)\}.$$

Suppose now that W is a finite-dimensional subspace of an inner product space V. By Theorem 2.6, given $x \in V$ we can express x uniquely in the form $x = a+b$ where $a \in W$ and $b \in W^\perp$. Then, by orthogonality,

$$\|x\|^2 = \langle a+b\,|\,a+b\rangle = \|a\|^2 + \|b\|^2.$$

It follows that, for every $y \in W$,

$$\begin{aligned}
\|x-y\|^2 = \|a-y+b\|^2 &= \|a-y\|^2 + \|b\|^2 + 2\langle a-y\,|\,b\rangle \\
&= \|a-y\|^2 + \|b\|^2 \\
&= \|a-y\|^2 + \|x-a\|^2 \\
&\geqslant \|x-a\|^2.
\end{aligned}$$

Thus we see that the element of W that is 'nearest' the element of x of V is the component a of x in W.

Now let $\{e_1,\ldots,e_n\}$ be an orthonormal basis of W. Express the element $a \in W$ that is nearest a given $x \in V$ as the linear combination

$$a = \sum_{i=1}^{n} \lambda_i e_i.$$

By Theorem 1.3, we have $\lambda_i = \langle a\,|\,e_i\rangle$ and by orthogonality

$$\langle x\,|\,e_i\rangle = \langle a+b\,|\,e_i\rangle = \langle a\,|\,e_i\rangle.$$

Thus we see that the element of W that is nearest to x is $\sum_{i=1}^{n}\langle x\,|\,e_i\rangle e_i$, the scalars being the Fourier coefficients.

Example 2.9

Let us apply the above observations to the inner product space V of continuous functions $f : [0,2\pi] \to \mathbb{R}$ under the inner product

$$\langle f\,|\,g\rangle = \int_0^{2\pi} fg.$$

An orthonormal subset of V is

$$S = \left\{x \mapsto \tfrac{1}{\sqrt{2\pi}},\ x \mapsto \tfrac{1}{\sqrt{\pi}}\cos kx,\ x \mapsto \tfrac{1}{\sqrt{\pi}}\sin kx\ ;\ k = 1,2,3,\ldots\right\}.$$

Let W_n be the subspace, of dimension $2n + 1$, with orthonormal basis

$$B_n = \{x \mapsto \tfrac{1}{\sqrt{2\pi}},\ x \mapsto \tfrac{1}{\sqrt{\pi}} \cos kx,\ x \mapsto \tfrac{1}{\sqrt{\pi}} \sin kx\ ;\ k = 1,\dots n\}.$$

Then the element f_n of W_n that is nearest a given $f \in V$ (or, put another way, is the best approximation to f in W_n) is

$$f_n(x) = c_0 \tfrac{1}{\sqrt{2\pi}} + \sum_{k=1}^{n}(c_k \cos kx + d_k \sin kx)$$

where $c_0 = \langle f(x) \mid \tfrac{1}{\sqrt{2\pi}}\rangle$ and

$$(\forall k \geqslant 1) \quad c_k = \langle f(x) \mid \tfrac{1}{\sqrt{\pi}} \cos kx\rangle, \quad d_k = \langle f(x) \mid \tfrac{1}{\sqrt{\pi}} \sin kx\rangle.$$

This can be written in the standard form

$$f_n(x) = \tfrac{1}{2}a_0 + \sum_{k=1}^{n}(a_k \cos kx + b_k \sin kx)$$

where $a_0 = \tfrac{1}{\pi} \int_0^{2\pi} f(x)\,dx$ and

$$(k \geqslant 1) \quad a_k = \tfrac{1}{\pi} \int_0^{2\pi} f(x) \cos kx\,dx; \quad b_k = \tfrac{1}{\pi} \int_0^{2\pi} f(x) \sin kx\,dx.$$

If f is infinitely differentiable then it can be shown that the sequence $(f_n)_{n \geqslant 1}$ is a Cauchy sequence having f as its limit. Thus we can write

$$f = \tfrac{1}{2}a_0 + \sum_{k \geqslant 1}(a_k \cos kx + b_k \sin kx)$$

which is the **Fourier series representation** of f.

Primary Decomposition

The characterisation of direct sums given in Theorem 2.5 opens the door to a deep study of linear mappings on finite-dimensional vector spaces and their representation by matrices. In order to embark on this, we require the following notion.

Definition

If V is a vector space over a field F and if $f : V \to V$ is linear then a subspace W of V is said to be *f*-**invariant** (or *f*-**stable**) if it satisfies the property $f^{\to}(W) \subseteq W$, i.e. if $x \in W \Rightarrow f(x) \in W$.

Example 3.1

If $f : V \to V$ is linear then $\operatorname{Im} f$ and $\operatorname{Ker} f$ are f-invariant subspaces of V.

Example 3.2

Let D be the differentiation map on the vector space $\mathbb{R}[X]$ of all real polynomials. Then D is linear and the subspace $\mathbb{R}_n[X]$ consisting of the polynomials of degree at most n is D-invariant.

Example 3.3

If $f : V \to V$ is linear and $x \in V$ with $x \neq 0_V$ then the subspace spanned by $\{x\}$ is f-invariant if and only if x is an eigenvector of f. In fact, the subspace spanned by $\{x\}$ is $Fx = \{\lambda x \; ; \; \lambda \in F\}$ and this is f-invariant if and only if for every $\lambda \in F$ there exists $\mu \in F$ such that $f(\lambda x) = \mu x$. Taking $\lambda = 1_F$ we see that x is an eigenvector of f. Conversely, if x is an eigenvector of f then $f(x) = \mu x$ for some $\mu \in F$ and so, for every $\lambda \in F$, we have $f(\lambda x) = \lambda f(x) = \lambda \mu x \in Fx$.

Example 3.4

Let $f : V \to V$ be linear. Then for every $p \in F[X]$ the subspace $\operatorname{Ker} p(f)$ of V is f-invariant.

EXERCISES

3.1 Let $f : \mathbb{R}^4 \to \mathbb{R}^4$ be given by
$$f(a,b,c,d) = (a + b + 2c - d,\ b + d,\ b + c,\ 2b - d).$$
Show that the subspace $W = \{x, 0, z, 0)\ ;\ x, z \in \mathbb{R}\}$ is f-invariant.

3.2 Let V be the vector space of real continuous functions defined on the interval $[0, 1]$. Let $\varphi : V \to V$ be given by $f \mapsto \varphi(f)$ where
$$(\forall x \in V) \qquad [\varphi(f)](x) = x \int_0^1 f.$$
Let W be the subspace consisting of those $f \in V$ of the form $f(x) = ax + b$ for some $a, b \in \mathbb{R}$. Prove that W is φ-invariant.

3.3 Let V be a finite-dimensional vector space and let p be a projection. If $f : V \to V$ is linear, prove that

  (1) $\operatorname{Im} p$ is f-invariant if and only if $p \circ f \circ p = f \circ p$;

  (2) $\operatorname{Ker} p$ is f-invariant if and only if $p \circ f \circ p = p \circ f$.

  Deduce that $\operatorname{Im} p$ and $\operatorname{Ker} p$ are f-invariant if and only if p and f commute.

3.4 Let V be a finite-dimensional vector space and let $f, g : V \to V$ be linear mappings such that $f \circ g = \operatorname{id}_V$. Prove that $g \circ f = \operatorname{id}_V$. Prove also that a subspace of V is f-invariant if and only if it is g-invariant. Does this hold if V is infinite-dimensional?

3.5 Let V be a finite-dimensional vector space and let $f : V \to V$ be linear. Suppose that $V = \bigoplus_{i=1}^{n} V_i$ where each V_i is an f-invariant subspace of V. If $f_i : V_i \to V$ is the restriction of f to V_i, prove that

  (a) $\operatorname{Im} f = \bigoplus_{i=1}^{n} \operatorname{Im} f_i$;

  (b) $\operatorname{Ker} f = \bigoplus_{i=1}^{n} \operatorname{Ker} f_i$.

Theorem 3.1

If $f : V \to V$ is linear then for every polynomial $p \in F[X]$ the subspaces $\operatorname{Im} p(f)$ and $\operatorname{Ker} p(f)$ are f-invariant.

Proof

Observe first that for every polynomial p we have $f \circ p(f) = p(f) \circ f$. It follows from this that if $x = p(f)(y)$ then $f(x) = p(f)[f(y)]$, whence $\operatorname{Im} p(f)$ is f-invariant. Likewise, if $p(f)(x) = 0_V$ then $p(f)[f(x)] = 0_V$, whence $\operatorname{Ker} p(f)$ is f-invariant (c.f. Exercise 3.3). $\square$

In what follows we shall often have occasion to deal with expressions of the form $p(f)$ where p is a polynomial and f is a linear mapping, and in so doing we shall find it convenient to denote composites by simple juxtaposition. Thus, for example, we shall write $fp(f)$ for $f \circ p(f)$.

Suppose now that V is of finite dimension n, that $f : V \to V$ is linear, and that the subspace W of V is f-invariant. Then f induces a linear mapping $\bar{f} : W \to W$, namely that given by the assignment $w \mapsto \bar{f}(w) = f(w) \in W$. Choose a basis $\{w_1, \ldots, w_r\}$ of W and extend it to a basis

$$\{w_1, \ldots, w_r, v_1, \ldots, v_{n-r}\}$$

of V. Consider the matrix of f relative to this basis. Since W is f-invariant each $f(w_i) \in W$ and so, for each i,

$$f(w_i) = \lambda_{i1}w_1 + \cdots + \lambda_{ir}w_r + 0\,v_1 + \cdots + 0\,v_{n-r},$$

whence it follows that this matrix is of the form

$$\begin{bmatrix} A & B \\ 0 & C \end{bmatrix}$$

where A is the $r \times r$ matrix that represents the mapping induced on W by f.

Suppose now that $V = V_1 \oplus V_2$ where both V_1 and V_2 are f-invariant subspaces of V. If B_1 is a basis of V_1 and B_2 is a basis of V_2 then by Theorem 2.2 we have that $B = B_1 \cup B_2$ is a basis of V, and it is readily seen that the matrix of f relative to B is of the form

$$\begin{bmatrix} A_1 & 0 \\ 0 & A_2 \end{bmatrix}$$

where A_1, A_2 represent the mappings induced on V_1, V_2 by f.

More generally, if $V = \bigoplus_{i=1}^{k} V_i$ where each V_i is f-invariant and if B_i is a basis of V_i for each i then the matrix of f relative to the basis $B = \bigcup_{i=1}^{k} B_i$ is of the **block diagonal form**

$$\begin{bmatrix} A_1 & & & \\ & A_2 & & \\ & & \ddots & \\ & & & A_k \end{bmatrix}$$

in which A_i is the matrix representing the mapping induced on V_i by f, so that A_i is of size $\dim V_i \times \dim V_i$.

Our objective now is to use the notions of direct sum and invariant subspace in order to find a basis of V such that the matrix of f relative to this basis has a particularly useful form. The key to this study is the following fundamental result.

Theorem 3.2

[Primary Decomposition Theorem] *Let V be a non-zero finite-dimensional vector space over a field F and let $f : V \to V$ be linear. Let the characteristic and minimum polynomials of f be*

$$c_f = p_1^{d_1} p_2^{d_2} \cdots p_k^{d_k}, \quad m_f = p_1^{e_1} p_2^{e_2} \cdots p_k^{e_k}$$

respectively, where $p_1, \ldots, p_k$ are distinct irreducible polynomials in $F[X]$. Then each of the subspaces $V_i = \operatorname{Ker} p_i^{e_i}(f)$ is f-invariant and $V = \bigoplus_{i=1}^{k} V_i$.

Moreover, if $f_i : V_i \to V_i$ is the linear mapping that is induced on V_i by f then the characteristic polynomial of f_i is $p_i^{d_i}$ and the minimum polynomial of f_i is $p_i^{e_i}$.

Proof

If $k = 1$ then $p_1^{e_1}(f) = m_f(f) = 0$ and trivially $V = \operatorname{Ker} p_1^{e_1}(f) = V_1$. Suppose then that $k \geqslant 2$. For $i = 1, \ldots, k$ let

$$q_i = m_f / p_i^{e_i} = \prod_{j \neq i} p_j^{e_j}.$$

Then there is no irreducible factor that is common to each of $q_1, \ldots, q_k$ and so there exist $a_1, \ldots, a_k \in F[X]$ such that

$$a_1 q_1 + a_2 q_2 + \cdots + a_k q_k = 1.$$

Writing $q_i a_i = t_i$ for each i, and substituting f in this polynomial identity, we obtain

$$(1) \qquad t_1(f) + \cdots + t_k(f) = \mathrm{id}_V.$$

Now by the definition of q_i we have that if $i \neq j$ then m_f divides $q_i q_j$. Consequently $q_i(f) q_j(f) = 0$ for $i \neq j$, and then

$$(2) \qquad (i \neq j) \qquad t_i(f) t_j(f) = 0.$$

By (1), (2), and Theorem 2.5 we see that each $t_i(f)$ is a projection and

$$V = \bigoplus_{i=1}^{k} \operatorname{Im} t_i(f).$$

Moreover, by Theorem 3.1, each of the subspaces $\operatorname{Im} t_i(f)$ is f-invariant.

We now show that

$$\operatorname{Im} t_i(f) = \operatorname{Ker} p_i^{e_i}(f).$$

Observe first that, since $p_i^{e_i} q_i = m_f$, we have $p_i^{e_i}(f) q_i(f) = m_f(f) = 0$ and therefore $p_i^{e_i}(f) q_i(f) a_i(f) = 0$ whence $\operatorname{Im} t_i(f) \subseteq \operatorname{Ker} p_i^{e_i}(f)$. To obtain the reverse inclusion, observe that for every j we have

$$t_j(f) = a_j(f) q_j(f) = a_j(f) \prod_{i \neq j} p_i^{e_i}(f),$$

from which we see that

$$\operatorname{Ker} p_i^{e_i}(f) \subseteq \bigcap_{j \neq i} \operatorname{Ker} t_j(f)$$

$$\subseteq \operatorname{Ker} \sum_{j \neq i} t_j(f)$$

$$= \operatorname{Ker} \left(\operatorname{id}_V - t_i(f) \right) \quad \text{by (1)}$$

$$= \operatorname{Im} t_i(f) \quad \text{by the Corollary to Theorem 2.4.}$$

As for the induced mapping $f_i : V_i \to V_i$, let m_i be its minimum polynomial. Since $p_i^{e_i}(f)$ is the zero map on V_i, so also is $p_i^{e_i}(f_i)$. Consequently, we have that $m_{f_i} | p_i^{e_i}$. Thus $m_{f_i} | m_f$ and the m_{f_i} are relatively prime. Suppose now that $g \in F[X]$ is a multiple of m_{f_i} for every i. Then clearly $g(f_i)$ is the zero map on V_i. For every

$$x = \sum_{i=1}^{k} v_i \in \bigoplus_{i=1}^{k} V_i = V \text{ we then have}$$

$$g(f)(x) = \sum_{i=1}^{k} g(f)(v_i) = \sum_{i=1}^{k} g(f_i)(v_i) = 0_V$$

and so $g(f) = 0$ and consequently $m_f | g$. Thus we see that m_f is the least common multiple of $m_{f_1}, \ldots, m_{f_k}$. Since these polynomials are relatively prime, we then have

$m_f = \prod_{i=1}^{k} m_{f_i}$. But we know that $m_f = \prod_{i=1}^{k} p_i^{e_i}$, and that $m_{f_i} | p_i^{e_i}$. Since all of the polynomials in question are monic, it follows that $m_{f_i} = p_i^{e_i}$ for $i = 1, \ldots, k$.

Finally, using Theorem 2.2, we can paste together bases of the subspaces V_i to form a basis of V with respect to which, as seen above, the matrix of f is of the block diagonal form

$$M = \begin{bmatrix} A_1 & & & \\ & A_2 & & \\ & & \ddots & \\ & & & A_k \end{bmatrix}.$$

Since, from the theory of determinants,

$$\det(XI - M) = \prod_{i=1}^{k} \det(XI - A_i)$$

we see that $c_f = \prod_{i=1}^{k} c_{f_i}$. Now we know that $m_{f_i} = p_i^{e_i}$ and so, since m_{f_i} and c_{f_i} have

the same zeros, we must have $c_{f_i} = p_i^{r_i}$ some $r_i \geq p_i$. Thus $\prod_{i=1}^{k} p_i^{r_i} = c_f = \prod_{i=1}^{k} p_i^{d_i}$ from

which it follows that $r_i = d_i$ for $i = 1, \ldots, k$. $\quad \square$

Corollary 1

$(i = 1, \ldots, k) \quad \dim V_i = d_i \deg p_i$.

Proof

dim V_i is the degree of $c_{f_i} = p_i^{d_i}$. □

Corollary 2

If all the eigenvalues of f lie in the ground field F, so that

$$c_f = (X - \lambda_1)^{d_1}(X - \lambda_2)^{d_2} \cdots (X - \lambda_k)^{d_k};$$
$$m_f = (X - \lambda_1)^{e_1}(X - \lambda_2)^{e_2} \cdots (X - \lambda_k)^{e_k},$$

then $V_i = \mathrm{Ker}\,(f - \lambda_i \mathrm{id}_V)^{e_i}$ *is f-invariant, of dimension* d_i, *and* $V = \displaystyle\bigoplus_{i=1}^{k} V_i$. □

Example 3.5

Consider the linear mapping $f : \mathbb{R}^3 \to \mathbb{R}^3$ given by

$$f(x, y, z) = (-z, x + z, y + z).$$

Relative to the standard ordered basis of $\mathbb{R}^3$, the matrix of f is

$$A = \begin{bmatrix} 0 & 0 & -1 \\ 1 & 0 & 1 \\ 0 & 1 & 1 \end{bmatrix}.$$

It is readily seen from this that $c_f = m_f = (X+1)(X-1)^2$. By Corollary 2 of Theorem 3.2 we have

$$\mathbb{R}^3 = \mathrm{Ker}\,(f + \mathrm{id}) \oplus \mathrm{Ker}\,(f - \mathrm{id})^2$$

with $\mathrm{Ker}\,(f + \mathrm{id})$ of dimension 1 and $\mathrm{Ker}\,(f - \mathrm{id})^2$ of dimension 2. Now

$$(f + \mathrm{id})(x, y, z) = (x - z, x + y + z, y + 2z)$$

so a basis for $\mathrm{Ker}\,(f + \mathrm{id})$ is $\{(1, -2, 1)\}$. Also,

$$(f - \mathrm{id})^2(x, y, z) = (x - y + z, -2x + 2y - 2z, x - y + z)$$

so a basis for $\mathrm{Ker}\,(f - \mathrm{id})^2$ is $\{(0, 1, 1), (1, 1, 0)\}$. Thus a basis for $\mathbb{R}^3$ with respect to which the matrix of f is in block diagonal form is

$$B = \{(1, -2, 1), (0, 1, 1), (1, 1, 0)\}.$$

The transition matrix from B to the standard basis is

$$P = \begin{bmatrix} 1 & 0 & 1 \\ -2 & 1 & 1 \\ 1 & 1 & 0 \end{bmatrix}$$

and the block diagonal matrix that represents f relative to the basis B is

$$P^{-1}AP = \begin{bmatrix} -1 & & \\ & 2 & 1 \\ & -1 & 0 \end{bmatrix}.$$

Example 3.6

Consider the differential equation

$$(D^n + a_{n-1}D^{n-1} + \cdots + a_1 D + a_0)f = 0$$

with constant (complex) coefficients. Let V be the solution space, i.e. the set of all infinitely differentiable functions satisfying the equation. From the theory of differential equations we have that V is finite-dimensional with dim $V = n$. Consider the polynomial

$$m = X^n + a_{n-1}X^{n-1} + \cdots + a_1 X + a_0.$$

Over $\mathbb{C}$, this polynomial factorises as

$$m = (X - \alpha_1)^{e_1}(X - \alpha_2)^{e_2} \cdots (X - \alpha_k)^{e_k}.$$

Then $D : V \to V$ is linear and its minimum polynomial is m. By Corollary 2 of Theorem 3.2, V is the direct sum of the solution spaces V_i of the differential equations

$$(D - \alpha_i \mathrm{id})^{e_i} f = 0.$$

Now the solutions of $(D - \alpha\, \mathrm{id})^r f = 0$ can be determined using the fact that, by a simple inductive argument,

$$(D - \alpha\, \mathrm{id})^r f = e^{\alpha t} D^r (e^{-\alpha t} f).$$

Thus f is a solution if and only if $D^r(e^{-\alpha t}f) = 0$, which is the case if and only if $e^{-\alpha t}f$ is a polynomial of degree at most $r - 1$. A basis for the solution space of $(D - \alpha\, \mathrm{id})^r f = 0$ is then $\{e^{\alpha t}, te^{\alpha t}, \ldots, t^{r-1}e^{\alpha t}\}$.

EXERCISES

3.6 Consider the linear mapping $f : \mathbb{R}^3 \to \mathbb{R}^3$ given by

$$f(x, y, z) = (2x + y - z, -2x - y + 3z, z).$$

Find the minimum polynomial of f and deduce that

$$\mathbb{R}^3 = \mathrm{Ker} f \oplus \mathrm{Ker}\, (f - \mathrm{id})^2.$$

Find a block diagonal matrix that represents f.

3.7 Consider the linear mapping $f : \mathbb{R}^3 \to \mathbb{R}^3$ given by

$$f(x, y, z) = (x + y + z, x + y + z, x + y + z).$$

Determine the characteristic and minimum polynomials of f. Show that the matrices

$$M = \begin{bmatrix} 1 & 1 & 1 \\ 1 & 1 & 1 \\ 1 & 1 & 1 \end{bmatrix}, \quad N = \begin{bmatrix} 3 & 0 & 0 \\ 0 & 0 & 0 \\ 0 & 0 & 0 \end{bmatrix}$$

are similar.

It is natural to consider special cases of the Primary Decomposition Theorem. Here we shall look at the situation in which each of the irreducible factors p_i of m_f is linear and each $e_i = 1$, i.e. when

$$m_f = (X - \lambda_1)(X - \lambda_2) \cdots (X - \lambda_k).$$

This gives the following important result, for which we recall that $f : V \to V$ is said to be **diagonalisable** if there is a basis of V consisting of eigenvectors of f; equivalently, if there is a basis of V with respect to which the matrix of f is diagonal.

Theorem 3.3

Let V be a non-zero finite-dimensional vector space and let $f : V \to V$ be linear. Then the following statements are equivalent:

(1) *the minimum polynomial m_f of f is a product of distinct linear factors;*
(2) *f is diagonalisable.*

Proof

$(1) \Rightarrow (2)$: Suppose that (1) holds and that

$$m_f = (X - \lambda_1)(X - \lambda_2) \cdots (X - \lambda_k)$$

where $\lambda_1, \ldots, \lambda_k$ are distinct elements of the ground field. By Theorem 3.2, V is the direct sum of the f-invariant subspaces $V_i = \mathrm{Ker}\,(f - \lambda_i \mathrm{id}_V)$. For every $x \in V_i$ we have $(f - \lambda_i \mathrm{id}_V)(x) = 0_V$ and so $f(x) = \lambda_i x$. Thus each λ_i is an eigenvalue of f, and every non-zero element of V_i is an eigenvector of f associated with λ_i. By Theorem 2.2 we can paste together bases of $V_1, \ldots, V_k$ to form a basis for V. Then V has a basis consisting of eigenvectors of f and so f is diagonalisable.

$(2) \Rightarrow (1)$: Suppose now that (2) holds and let $\lambda_1, \ldots, \lambda_k$ be distinct eigenvalues of f. Consider the polynomial

$$p = (X - \lambda_1)(X - \lambda_2) \cdots (X - \lambda_k).$$

Clearly, $p(f)$ maps every basis vector to 0_V and consequently $p(f) = 0$. The minimum polynomial m_f therefore divides p, and must coincide with p since every eigenvalue of f is a zero of m_f. $\square$

Example 3.7

Consider the linear mapping $f : \mathbb{R}^3 \to \mathbb{R}^3$ given by

$$f(x, y, z) = (7x - y - 2z, \ -x + 7y + 2z, \ -2x + 2y + 10z).$$

Relative to the standard ordered basis of $\mathbb{R}^3$, the matrix of f is

$$A = \begin{bmatrix} 7 & -1 & -2 \\ -1 & 7 & 2 \\ -2 & 2 & 10 \end{bmatrix}.$$

The reader can verify that $c_f = (X-6)^2(X-12)$ and $m_f = (X-6)(X-12)$. It follows by Theorem 3.3 that f is diagonalisable.

EXERCISES

3.8 If $f : \mathbb{R}^n \to \mathbb{R}^n$ is linear and such that $f^3 = f$, prove that f is diagonalisable.

3.9 Let $f : \mathbb{R}^3 \to \mathbb{R}^3$ be the linear mapping given by the prescription

$$f(x, y, z) = (-2x - y + z, \; 2x + y - 3z, \; -z).$$

Find the eigenvalues and the minimum polynomial of f and show that f is not diagonalisable.

3.10 Determine whether or not each of the following mappings $f : \mathbb{R}^3 \to \mathbb{R}^3$ is diagonalisable:

(a) $f(x, y, z) = (3x - y + z, \; -x + 5y - z, \; x - y + 3z)$;

(b) $f(x, y, z) = (2x, \; x + 2y, \; -x + y + z)$;

(c) $f(x, y, z) = (x - z, \; 2y, \; x + y + 3z)$.

An interesting result concerning diagonalisable mappings that will be useful to us later is the following:

Theorem 3.4

Let V be a non-zero finite-dimensional vector space and let $f, g : V \to V$ be diagonalisable linear mappings. Then f and g are simultaneously diagonalisable (in the sense that there is a basis of V that consists of eigenvectors of both f and g) if and only if $f \circ g = g \circ f$.

Proof

$\Rightarrow$: Suppose that there is a basis $\{v_1, \ldots, v_n\}$ of V such that each v_i is an eigenvector of both f and g. If $f(v_i) = \lambda_i v_i$ and $g(v_i) = \mu_i v_i$ then we have

$$f[g(v_i)] = \lambda_i \mu_i v_i = \mu_i \lambda_i v_i = g[f(v_i)].$$

Thus $f \circ g$ and $g \circ f$ agree on a basis and so $f \circ g = g \circ f$.

$\Leftarrow$: Suppose now that $f \circ g = g \circ f$. Since f is diagonalisable its minimum polynomial is of the form

$$m_f = (X - \lambda_1)(X - \lambda_2) \cdots (X - \lambda_k)$$

where $\lambda_1, \ldots, \lambda_k$ are distinct eigenvalues of f. By Corollary 2 of Theorem 3.2, we have $V = \bigoplus_{i=1}^{k} V_i$ where $V_i = \mathrm{Ker}\,(f - \lambda_i \mathrm{id}_V)$. Now since, by the hypothesis, $f \circ g = g \circ f$ we have, for $v_i \in V_i$,

$$f[g(v_i)] = g[f(v_i)] = g(\lambda_i v_i) = \lambda_i g(v_i)$$

and so $g(v_i) \in V_i$. Thus each V_i is g-invariant. Now let $g_i : V_i \rightarrow V_i$ be the linear mapping thus induced by g. Since g is diagonalisable, so is each g_i, for the minimum polynomial of g_i divides that of g. We can therefore find a basis B_i of V_i consisting of eigenvectors of g_i. Since every eigenvector of g_i is an eigenvector of g, and since every element of V_i is an eigenvector of f, it follows that $\bigcup\limits_{i=1}^{k} B_i$ is a basis of V that consists of eigenvectors of both f and g. $\square$

Corollary

Let A, B be $n \times n$ matrices over a field F. If A and B are diagonalisable then they are simultaneously diagonalisable (i.e. there is an invertible matrix P such that $P^{-1}AP$ and $P^{-1}BP$ are diagonal) if and only if $AB = BA$. $\square$

Reduction to Triangular Form

The Primary Decomposition Theorem shows that for a linear mapping f on a finite-dimensional vector space V there is a basis of V with respect to which f can be represented by a block diagonal matrix. As we have seen, in the special situation where the minimum polynomial of f is a product of distinct linear factors, this matrix is diagonal. We now turn our attention to a slightly more general situation, namely that in which the minimum polynomial of f factorises as a product of linear factors that are not necessarily distinct, i.e. is of the form

$$m_f = \prod_{i=1}^{k} (X - \lambda_i)^{e_i}$$

where each $e_i \geqslant 1$. This, of course, is always the case when the ground field is $\mathbb{C}$, so the results we shall establish will be valid for all linear mappings on a finite-dimensional complex vector space. To let the cat out of the bag, our specific objective is to show that when the minimum polynomial of f factorises completely there is a basis of V with respect to which the matrix of f is **triangular**. We recall that a matrix $A = [a_{ij}]_{n \times n}$ is (upper) triangular if $a_{ij} = 0$ whenever $i > j$.

In order to see how to proceed, we observe first that by Corollary 2 of Theorem 3.2 we can write V as a direct sum of the f-invariant subspaces $V_i = \mathrm{Ker}\,(f - \lambda_i \mathrm{id}_V)^{e_i}$. Let $f_i : V_i \to V_i$ be the linear mapping induced on the 'primary component' V_i by f, and consider the mapping $f_i - \lambda_i \mathrm{id}_{V_i} : V_i \to V_i$. We have that $(f_i - \lambda_i \mathrm{id}_{V_i})^{e_i}$ is the zero map on V_i, and so $f_i - \lambda_i \mathrm{id}_{V_i}$ is **nilpotent,** in the following sense.

Definition

A linear mapping $f : V \to V$ is said to be **nilpotent** if $f^m = 0$ for some positive integer m. Likewise, a square matrix A is said to be **nilpotent** if there is a positive integer m such that $A^m = 0$.

Example 4.1

The linear mapping $f : \mathbb{R}^3 \to \mathbb{R}^3$ given by $f(x, y, z) = (0, x, y)$ is nilpotent. In fact, $f^2(x, y, z) = f(0, x, y) = (0, 0, x)$ and then $f^3 = 0$.

Example 4.2

If $f : \mathbb{C}^n \to \mathbb{C}^n$ is such that all the eigenvalues of f are 0 then we have $c_f = X^n$. By the Cayley–Hamilton Theorem, $f^n = c_f(f) = 0$ so that f is nilpotent.

Example 4.3

The differentiation mapping $D : \mathbb{R}_n[X] \to \mathbb{R}_n[X]$ is nilpotent.

EXERCISES

4.1 Show that the linear mapping $f : \mathbb{R}^3 \to \mathbb{R}^3$ defined by

$$f(x, y, z) = (-x - y - z,\ 0,\ x + y + z)$$

is nilpotent.

4.2 If $f : \mathbb{R}_2[X] \to \mathbb{R}_2[X]$ is the linear mapping whose action on the basis $\{1, X, X^2\}$ of $\mathbb{R}_2[X]$ is given by

$$f(1) = -5 - 8X - 5X^2$$

$$f(X) = 1 + X + X^2$$

$$f(X^2) = 4 + 7X + 4X^2$$

show that f is nilpotent.

4.3 If $f : V \to V$ is nilpotent prove that the only eigenvalue of f is 0.

4.4 If $V = \mathrm{Mat}_{n \times n} \mathbb{R}$ and $A \in V$, show that $f_A : V \to V$ given by

$$f_A(X) = AX - XA$$

is linear. Prove that if A is nilpotent then so is f_A.

As we shall see, the notion of a nilpotent mapping shares an important role with with that of a diagonalisable mapping. In order to prepare the way for this we now produce, relative to a nilpotent linear mapping, a particularly useful basis.

Theorem 4.1

Let V be a non-zero finite-dimensional vector space and let $f : V \to V$ be a nilpotent linear mapping. Then there is an ordered basis $\{v_1, \ldots, v_n\}$ of V such that

$$f(v_1) = 0_V;$$

$$f(v_2) \in \mathrm{Span}\{v_1\};$$

$$f(v_3) \in \mathrm{Span}\{v_1, v_2\};$$

$$\vdots$$

$$f(v_n) \in \mathrm{Span}\{v_1, \ldots, v_{n-1}\}.$$

Proof

Since f is nilpotent there is a positive integer m such that $f^m = 0$. If $f = 0$ then the situation is trivial, for every basis of V satisfies the stated conditions. So we can assume that $f \neq 0$. Let k be the smallest positive integer such that $f^k = 0$. Then $f^i \neq 0$ for $1 \leqslant i \leqslant k - 1$. Since $f^{k-1} \neq 0$ there exists $v \in V$ such that $f^{k-1}(v) \neq 0_V$. Let $v_1 = f^{k-1}(v)$ and observe that $f(v_1) = 0_V$. We now proceed recursively. Suppose that we have been able to find $v_1, \ldots, v_r$ satisfying the conditions and consider the subspace $W = \mathrm{Span}\{v_1, \ldots, v_r\}$. If $W = V$ then there is nothing more to prove. If $W \neq V$ there are two possibilities, depending on whether $\mathrm{Im} f \subseteq W$ or $\mathrm{Im} f \nsubseteq W$. In the former case, let v_{r+1} be any element of $V \backslash W$. In the latter case, since we have the chain

$$\{0_V\} = \mathrm{Im} f^k \subseteq \mathrm{Im} f^{k-1} \subseteq \cdots \subseteq \mathrm{Im} f^2 \subseteq \mathrm{Im} f,$$

there is a positive integer j such that $\mathrm{Im} f^j \nsubseteq W$ and $\mathrm{Im} f^{j+1} \subseteq W$. In this case we choose $v_{r+1} \in \mathrm{Im} f^j$ with $v_{r+1} \notin W$. Each of these choices is such that $\{v_1, \ldots, v_{r+1}\}$ is linearly independent, with $f(v_{r+1}) \in W$. $\square$

Corollary

If $f : V \to V$ is nilpotent then there is an ordered basis of V with respect to which the matrix of f is upper triangular with all diagonal entries 0.

Proof

From the above, the action of f on the basis $\{v_1, \ldots, v_n\}$ can be described by

$$f(v_1) = 0v_1 + 0v_2 + 0v_3 + \cdots + 0v_n$$
$$f(v_2) = a_{12}v_1 + 0v_2 + 0v_3 + \cdots + 0v_n$$
$$f(v_3) = a_{13}v_1 + a_{23}v_2 + 0v_3 + \cdots + 0v_n$$
$$\vdots$$
$$f(v_n) = a_{1n}v_1 + a_{2n}v_2 + a_{3n}v_3 + \cdots + 0v_n$$

where the a_{ij} belong to the ground field.

The matrix of f relative to this basis is then the $n \times n$ matrix

$$A = \begin{bmatrix} 0 & a_{12} & a_{13} & \cdots & a_{1n} \\ 0 & 0 & a_{23} & \cdots & a_{2n} \\ \vdots & \vdots & \vdots & & \vdots \\ 0 & 0 & 0 & \cdots & a_{n-1,n} \\ 0 & 0 & 0 & \cdots & 0 \end{bmatrix}$$

which is upper triangular with all diagonal entries 0. $\square$

Returning to our consideration of the primary components, we can apply the above results to the nilpotent linear mapping $g_i = f_i - \lambda_i \mathrm{id}_{V_i}$ on the direct summand V_i of dimension d_i. Since

$$f_i = g_i + \lambda_i \mathrm{id}_{V_i},$$

we deduce from the above that there is an ordered basis of the subspace V_i with respect to which

$$\mathrm{Mat}\, f_i = \mathrm{Mat}\, g_i + \lambda_i \, \mathrm{Mat}\, \mathrm{id}_{V_i}$$

$$= \begin{bmatrix} \lambda_i & a_{12} & a_{13} & \cdots & a_{1n} \\ 0 & \lambda_i & a_{23} & \cdots & a_{2n} \\ \vdots & \vdots & \vdots & & \vdots \\ 0 & 0 & 0 & \cdots & a_{n-1,n} \\ 0 & 0 & 0 & \cdots & \lambda_i \end{bmatrix}.$$

Consequently, we have the following result.

Theorem 4.2

[Triangular Form]   *Let V be a non-zero finite-dimensional vector space over a field F and let $f : V \to V$ be a linear mapping whose characteristic and minimum polynomials are*

$$c_f = \prod_{i=1}^{k}(X - \lambda_i)^{d_i}, \quad m_f = \prod_{i=1}^{k}(X - \lambda_i)^{e_i}$$

for distinct $\lambda_1, \ldots, \lambda_k \in F$ and $e_i \leqslant d_i$. Then there is an ordered basis of V with respect to which the matrix of f is upper triangular; more specifically, is a block diagonal matrix

$$M = \begin{bmatrix} A_1 & & & \\ & A_2 & & \\ & & \ddots & \\ & & & A_k \end{bmatrix}$$

in which A_i is a $d_i \times d_i$ upper triangular matrix

$$\begin{bmatrix} \lambda_i & ? & \cdots & ? \\ 0 & \lambda_i & \cdots & ? \\ \vdots & \vdots & & \vdots \\ 0 & 0 & \cdots & \lambda_i \end{bmatrix}$$

in which the entries marked ? are elements of F. $\square$

Corollary

Every square matrix over the field $\mathbb{C}$ of complex numbers is similar to an upper triangular matrix. $\square$

Example 4.4

Consider the linear mapping $f : \mathbb{R}^3 \to \mathbb{R}^3$ given by

$$f(x, y, z) = (x + z, \; 2y + z, \; -x + 3z).$$

Relative to the standard ordered basis of $\mathbb{R}^3$, the matrix of f is

$$A = \begin{bmatrix} 1 & 0 & 1 \\ 0 & 2 & 1 \\ -1 & 0 & 3 \end{bmatrix}.$$

The reader will readily verify that $c_f = m_f = (X - 2)^3$. It follows by Corollary 2 of Theorem 3.2 that

$$\mathbb{R}^3 = \mathrm{Ker}\,(f - 2\,\mathrm{id})^3.$$

We now find a basis for $\mathbb{R}^3$ in the style of Theorem 4.1. First, we note that

$$(f - 2\,\mathrm{id})(x, y, z) = (-x + z, \; z, \; -x + z).$$

We therefore choose, for example,

$$v_1 = (0, 1, 0) \in \mathrm{Ker}\,(f - 2\,\mathrm{id}) \backslash \{0\}.$$

As for v_2, we require that v_2 be independent of v_1 and such that

$$(f - 2\,\mathrm{id})(v_2) \in \mathrm{Span}\,\{v_1\};$$

i.e. we have to choose $v_2 = (x, y, z)$ independent of $v_1 = (0, 1, 0)$ such that

$$(-x + z, \; z, \; -x + z) = \alpha(0, 1, 0).$$

We may take, for example, $\alpha = 1$ and choose $v_2 = (1, 0, 1)$.

We now require v_3 independent of $\{v_1, v_2\}$ such that

$$(f - 2\,\mathrm{id})(v_3) \in \mathrm{Span}\,\{v_1, v_2\}.$$

We may choose, for example $v_3 = (0, 0, 1)$. Consider now the basis

$$B = \{v_1, v_2, v_3\} = \{(0, 1, 0), (1, 0, 1), (0, 0, 1)\}.$$

The transition matrix from B to the standard basis of $\mathbb{R}^3$ is

$$P = \begin{bmatrix} 0 & 1 & 0 \\ 1 & 0 & 0 \\ 0 & 1 & 1 \end{bmatrix}$$

and so the matrix of f relative to B is the upper triangular matrix

$$P^{-1}AP = \begin{bmatrix} 2 & 1 & 1 \\ 0 & 2 & 1 \\ 0 & 0 & 2 \end{bmatrix}.$$

Example 4.5

Referring to Example 3.5, consider again the linear mapping $f : \mathbb{R}^3 \to \mathbb{R}^3$ given by

$$f(x, y, z) = (-z, \; x + z, \; y + z).$$

We have

$$\mathbb{R}^3 = \operatorname{Ker}(f + \mathrm{id}) \oplus \operatorname{Ker}(f - \mathrm{id})^2$$

with $\operatorname{Ker}(f + \mathrm{id})$ of dimension 1 and $\operatorname{Ker}(f - \mathrm{id})^2$ of dimension 2.

Since

$$(f + \mathrm{id})(x, y, z) = (x - z, \; x + y + z, \; y + 2z),$$

a basis for $V_1 = \operatorname{Ker}(f + \mathrm{id})$ is $\{(1, -2, 1)\}$.

Now consider finding a basis for $V_2 = \operatorname{Ker}(f - \mathrm{id})^2$ in the style of Theorem 4.1. First note that

$$(f - \mathrm{id})(x, y, z) = (-x - z, \; x - y + z, \; y).$$

Begin by choosing

$$w_1 = (-1, 0, 1) \in \operatorname{Ker}(f - \mathrm{id}) \backslash \{0\}.$$

We now require w_2 independent of w_1 with

$$(f - \mathrm{id})(w_2) \in \operatorname{Span}\{w_1\};$$

i.e. we have to find $w_2 = (x, y, z)$ independent of $w_1 = (-1, 0, 1)$ such that

$$(-x - z, \; x - y + z, \; y) = \alpha(-1, 0, 1).$$

We may take, for example, $\alpha = 1$ and choose $w_2 = (0, 1, 1)$. Now since

$$(f - \mathrm{id})(w_1) = (0, 0, 0) = 0w_1 + 0w_2$$

$$(f - \mathrm{id})(w_2) = (-1, 0, 1) = 1w_1 + 0w_2$$

we see that the matrix of $f - \mathrm{id}$ relative to the basis $\{w_1, w_2\}$ is

$$\begin{bmatrix} 0 & 1 \\ 0 & 0 \end{bmatrix}.$$

The matrix of the mapping f_2 that is induced on $V_2 = \operatorname{Ker}(f - \mathrm{id})^2$ by f is then

$$\begin{bmatrix} 1 & 1 \\ 0 & 1 \end{bmatrix}.$$

Consequently, the matrix of f relative to the basis

$$B = \{(1, -2, 1), (-1, 0, 1), (0, 1, 1)\}$$

is the upper triangular matrix

$$\begin{bmatrix} -1 & 0 & 0 \\ 0 & 1 & 1 \\ 0 & 0 & 1 \end{bmatrix}.$$

EXERCISES

4.5 Let $f : \mathbb{R}^3 \to \mathbb{R}^3$ be the linear mapping given by

$$f(x, y, z) = (2x + y - z, -2x - y + 3z, z).$$

Find the characteristic and minimum polynomials of f and show that f is not diagonalisable. Find an ordered basis of $\mathbb{R}^3$ with respect to which the matrix of f is upper triangular.

4.6 Let $f : \mathbb{R}^3 \to \mathbb{R}^3$ be the linear mapping given by

$$f(x, y, z) = (2x - 2y, x - y, -x + 3y + z).$$

Find the characteristic and minimum polynomials of f and show that f is not diagonalisable. Find an ordered basis of $\mathbb{R}^3$ with respect to which the matrix of f is upper triangular.

We have seen above that if $f : V \to V$ is linear and every eigenvalue of f lies in the ground field of V then each induced mapping f_i on the f-invariant subspace $V_i = \mathrm{Ker}\,(f - \lambda_i \mathrm{id}_V)^{e_i}$ can be written in the form $f_i = g_i + \lambda_i \mathrm{id}_{V_i}$ where g_i is nilpotent. Clearly, $\lambda_i \mathrm{id}_{V_i}$ is diagonalisable (its minimum polynomial being $X - \lambda_i$). Thus every induced mapping f_i has a decomposition as the sum of a diagonalisable mapping and a nilpotent mapping that, moreover, commute. That this is true of f itself is the substance of the following important result.

Theorem 4.3

[Jordan Decomposition] *Let V be a non-zero finite-dimensional vector space over a field F and let $f : V \to V$ be a linear mapping all of whose eigenvalues belong to F. Then there is a diagonalisable linear mapping $\delta : V \to V$ and a nilpotent linear mapping $\eta : V \to V$ such that $f = \delta + \eta$ and $\delta \circ \eta = \eta \circ \delta$. Moreover, there are polynomials $p, q \in F[X]$ such that $\delta = p(f)$ and $\eta = q(f)$. Furthermore, δ and η are uniquely determined, in the sense that if $\delta', \eta' : V \to V$ are respectively diagonalisable and nilpotent linear mappings such that $f = \delta' + \eta'$ with $\delta' \circ \eta' = \eta' \circ \delta'$ then $\delta = \delta'$ and $\eta = \eta'$.*

Proof

The minimum polynomial of f is $m_f = \prod_{i=1}^{k} (X - \lambda_i)^{e_i}$ where $\lambda_1, \ldots, \lambda_k \in F$ are distinct. Moreover, $V = \bigoplus_{i=1}^{k} V_i$ where $V_i = \mathrm{Ker}\,(f - \lambda_i \mathrm{id}_V)^{e_i}$.

Let $\delta : V \to V$ be given by $\delta = \sum_{i=1}^{k} \lambda_i p_i$ where $p_i : V \to V$ is the projection on V_i parallel to $\sum_{j \neq i} V_j$. Then for every $v_i \in V_i$ we have $\delta(v_i) = \left(\sum_{j=1}^{k} \lambda_j p_j \right)(v_i) = \lambda_i v_i$ and consequently V has a basis consisting of eigenvectors of δ and so δ is diagonalisable.

Now define $\eta = f - \delta$. Then for every $v_i \in V_i$ we have

$$\eta(v_i) = f(v_i) - \delta(v_i) = (f - \lambda_i \mathrm{id}_V)(v_i)$$

and consequently $\eta^{e_i}(v_i) = (f - \lambda_i \mathrm{id}_V)^{e_i}(v_i) = 0_V$. It follows that, for some r, $\mathrm{Ker}\,\eta^r$ contains a basis of V, so $\eta^r = 0$ and hence η is nilpotent.

Now since $V = \bigoplus_{i=1}^{k} V_i$ every $v \in V$ can be written uniquely in the form $v = v_1 + \cdots + v_k$ with $v_i \in V_i$. Since each V_i is f-invariant, we then have

$$p_i[f(v)] = p_i[f(v_1) + \cdots + f(v_k)] = f(v_i) = f[p_i(v)]$$

and hence $p_i \circ f = f \circ p_i$ for each i. Consequently,

$$\delta \circ f = \sum_{i=1}^{k} \lambda_i p_i \circ f = \sum_{i=1}^{k} \lambda_i (p_i \circ f) = \sum_{i=1}^{k} \lambda_i (f \circ p_i) = f \circ \sum_{i=1}^{k} \lambda_i p_i = f \circ \delta.$$

It follows from this that

$$\delta \circ \eta = \delta(f - \delta) = \delta f - \delta^2 = f\delta - \delta^2 = (f - \delta)\delta = \eta \circ \delta.$$

We now show that there are polynomials $p, q \in F[X]$ such that $\delta = p(f)$ and $\eta = q(f)$. For this purpose, we observe first that $p_i = t_i(f)$ where t_i is the polynomial described in the proof of Theorem 3.2. Then, by the definition of δ, we have $\delta = p(f)$ where $p = \sum_{i=1}^{k} \lambda_i t_i$. Since $\eta = f - \delta$, there is then a polynomial q such that $\eta = q(f)$.

As for uniqueness, suppose that $\delta', \eta' : V \to V$ are diagonalisable and nilpotent respectively, with $f = \delta' + \eta'$ and $\delta' \circ \eta' = \eta' \circ \delta'$. Since, as we have just seen, there are polynomials p, q such that $\delta = p(f)$ and $\eta = q(f)$ it follows that $\delta' \circ \delta = \delta \circ \delta'$ and that $\eta' \circ \eta = \eta \circ \eta'$. Now since $\delta + \eta = f = \delta' + \eta'$ we have $\delta - \delta' = \eta' - \eta$ and so, since η, η' commute, we can use the binomial theorem to deduce from the fact that η and η' are nilpotent that so also is $\eta' - \eta$, which can therefore be represented by a nilpotent matrix N. Also, since δ, δ' commute it follows by Theorem 3.4 that there is a basis of V consisting of eigenvectors of both δ and δ'. Each such eigenvector is then an eigenvector of $\delta - \delta'$, and consequently $\delta - \delta'$ is represented by a diagonal matrix D. Now N and D are similar, and the only possibility is $N = D = 0$. Consequently, we have $\delta - \delta' = 0 = \eta' - \eta$ whence $\delta' = \delta$ and $\eta' = \eta$ as required. □

There is, of course, a corresponding result in terms of square matrices, namely that if all the eigenvalues of $A \in \mathrm{Mat}_{n \times n} F$ lie in F then A can be expressed uniquely as the sum of a diagonalisable matrix D and a nilpotent matrix N with $DN = ND$.

Example 4.6

In Example 4.5 we saw that if $f : \mathbb{R}^3 \to \mathbb{R}^3$ is given by

$$f(x, y, z) = (-z, x + z, y + z)$$

then relative to the basis

$$B = \{(1, -2, 1), (-1, 0, 1), (0, 1, 1)\}$$

the matrix of f is

$$T = \begin{bmatrix} -1 & 0 & 0 \\ 0 & 1 & 1 \\ 0 & 0 & 1 \end{bmatrix}.$$

Clearly, we can write $T = D + N$ where

$$D = \begin{bmatrix} -1 & 0 & 0 \\ 0 & 1 & 0 \\ 0 & 0 & 1 \end{bmatrix}, \quad N = \begin{bmatrix} 0 & 0 & 0 \\ 0 & 0 & 1 \\ 0 & 0 & 0 \end{bmatrix}$$

and this is the Jordan decomposition of T.

To refer matters back to the standard ordered basis, we compute PDP^{-1} and PNP^{-1} where P is the transition matrix

$$P = \begin{bmatrix} 1 & -1 & 0 \\ -2 & 0 & 1 \\ 1 & 1 & 1 \end{bmatrix}.$$

It is readily seen that

$$PDP^{-1} = \begin{bmatrix} \frac{1}{2} & \frac{1}{2} & -\frac{1}{2} \\ 1 & 0 & 1 \\ -\frac{1}{2} & \frac{1}{2} & \frac{1}{2} \end{bmatrix}, \quad PNP^{-1} = \begin{bmatrix} -\frac{1}{2} & -\frac{1}{2} & -\frac{1}{2} \\ 0 & 0 & 0 \\ \frac{1}{2} & \frac{1}{2} & \frac{1}{2} \end{bmatrix}.$$

Consequently, the diagonal part of f is given by

$$d_f(x, y, z) = (\tfrac{1}{2}x + \tfrac{1}{2}y - \tfrac{1}{2}z, \ x + z, \ -\tfrac{1}{2}x + \tfrac{1}{2}y + \tfrac{1}{2}z),$$

and the nilpotent part of f is given by

$$n_f(x, y, z) = (-\tfrac{1}{2}x - \tfrac{1}{2}y - \tfrac{1}{2}z, \ 0, \ \tfrac{1}{2}x + \tfrac{1}{2}y + \tfrac{1}{2}z).$$

EXERCISES

4.7 Determine the Jordan decomposition of the linear mappings described in Exercises 4.5 and 4.6.

4.8 Consider the linear mapping $f : \mathbb{R}^3 \to \mathbb{R}^3$ whose action on a basis $\{b_1, b_2, b_3\}$ of $\mathbb{R}^3$ is given by

$$f(b_1) = -b_1 + 2b_3;$$
$$f(b_2) = 3b_1 + 2b_2 + b_3;$$
$$f(b_3) = -b_3.$$

(a) Show that the minimum polynomial of f is $(X + 1)^2(X - 2)$.

(b) Determine a basis of $\mathbb{R}^3$ with respect to which the matrix of f is upper triangular.

(c) Find the Jordan decomposition of f.

The Jordan decomposition of a linear mapping f (or of a square matrix A) is particularly useful in computing powers of f (or of A). Indeed, since $f = \delta + \eta$ where δ, η commute, we can apply the binomial theorem to obtain

$$f^n = (\delta + \eta)^n = \sum_{i=1}^{n} \binom{n}{i} \delta^{n-i} \eta^i.$$

The powers of δ are easily computed (by considering the powers of the corresponding diagonal matrix), and all powers of η from some point on are zero.

EXERCISE

4.9 Determine the n-th power of the matrix

$$A = \begin{bmatrix} 0 & 0 & -1 \\ 1 & 0 & 1 \\ 0 & 1 & 1 \end{bmatrix}.$$

Reduction to Jordan Form

It is natural to ask if we can improve on the triangular form. In order to do so, it is clearly necessary to find 'better' bases for the subspaces that appear as the direct summands (or primary components) in the Primary Decomposition Theorem. So let us take a closer look at nilpotent mappings.

Definition

If the linear mapping $f : V \to V$ is nilpotent then by the **index** of f we shall mean the smallest positive integer k such that $f^k = 0$.

Example 5.1

As seen in Example 4.1, the mapping $f : \mathbb{R}^3 \to \mathbb{R}^3$ given by $f(x, y, z) = (0, x, y)$ is nilpotent. It is of index 3.

EXERCISES

5.1 Let $f : V \to V$ be linear and nilpotent of index p. Prove that if $x \in V$ is such that $f^{p-1}(x) \neq 0_V$ then

$$\{x, f(x), \ldots, f^{p-1}(x)\}$$

is a linearly independent subset of V.

5.2 Let V be a vector space of dimension n over a field F. If $f : V \to V$ is linear prove that f is nilpotent of index n if and only if there is an ordered basis of V with respect to which the matrix of f is

$$\begin{bmatrix} 0 & 0 \\ I_{n-1} & 0 \end{bmatrix}.$$

Deduce that an $n \times n$ matrix A over F is nilpotent of index n if and only if A is similar to this matrix.

In order to proceed, we observe the following simple facts.

Theorem 5.1

If $f : V \rightarrow V$ is linear then, for every positive integer i,
  (1) $\operatorname{Ker} f^i \subseteq \operatorname{Ker} f^{i+1}$;
  (2) *if $x \in \operatorname{Ker} f^{i+1}$ then $f(x) \in \operatorname{Ker} f^i$.*

Proof

  (1) If $x \in \operatorname{Ker} f^i$ then $f^i(x) = 0_V$ gives $f^{i+1}(x) = f[f^i(x)] = f(0_V) = 0_V$ and therefore $x \in \operatorname{Ker} f^{i+1}$.
  (2) If $x \in \operatorname{Ker} f^{i+1}$ then $f^i[f(x)] = f^{i+1}(x) = 0_V$ and so $f(x) \in \operatorname{Ker} f^i$. $\square$

In general, for a linear mapping $f : V \rightarrow V$ we have, by Theorem 5.1(1), the chain of subspaces

$$\{0_V\} \subseteq \operatorname{Ker} f \subseteq \operatorname{Ker} f^2 \subseteq \cdots \subseteq \operatorname{Ker} f^i \subseteq \operatorname{Ker} f^{i+1} \subseteq \cdots .$$

In the case where f is nilpotent, the following situation holds.

Theorem 5.2

Let V be a non-zero vector space over a field F and let $f : V \rightarrow V$ be a linear mapping that is nilpotent of index k. Then there is the chain of distinct subspaces

$$\{0_V\} \subset \operatorname{Ker} f \subset \operatorname{Ker} f^2 \subset \cdots \subset \operatorname{Ker} f^{k-1} \subset \operatorname{Ker} f^k = V.$$

Proof

Observe first that $\operatorname{Ker} f \neq \{0_V\}$, for otherwise from $f^k(x) = 0_V$ we would have $f^{k-1}(x) = 0_V$ for every x, and this contradicts the hypothesis that f is of index k.
  In view of Theorem 5.1, it now suffices to show that

$$(i = 1, \ldots, k - 1) \qquad \operatorname{Ker} f^i \neq \operatorname{Ker} f^{i+1}.$$

In fact, suppose that there exists $i \in \{1, \ldots, k - 1\}$ such that $\operatorname{Ker} f^i = \operatorname{Ker} f^{i+1}$. Then, for every $x \in V$, we have

$$0_V = f^k(x) = f^{i+1}[f^{k-(i+1)}(x)]$$

whence $f^{k-(i+1)}(x) \in \operatorname{Ker} f^{i+1} = \operatorname{Ker} f^i$ and so

$$0_V = f^i[f^{k-(i+1)}(x)] = f^{k-1}(x).$$

This produces the contradiction $f^{k-1} = 0$. $\square$

In connection with the above, the following result will prove to be very useful.

Theorem 5.3

Let V be a finite-dimensional vector space and let $f : V \rightarrow V$ be linear and such that $\operatorname{Ker} f^i \subset \operatorname{Ker} f^{i+1}$. If $\{v_1, \ldots, v_s\}$ is a basis of $\operatorname{Ker} f^i$ that is extended to a basis $\{v_1, \ldots, v_s, w_1, \ldots, w_t\}$ of $\operatorname{Ker} f^{i+1}$ then $S = \{f(w_1), \ldots, f(w_t)\}$ is a linearly independent subset of $\operatorname{Ker} f^i$.

Proof

If $x \in \operatorname{Ker} f^{i+1}$ then by Theorem 5.1(2) we have $f(x) \in \operatorname{Ker} f^{i}$. Thus we see that $S \subseteq \operatorname{Ker} f^{i}$. To see that S is linearly independent, suppose that

$$\mu_1 f(w_1) + \cdots + \mu_t f(w_t) = 0_V.$$

Then

$$\mu_1 w_1 + \cdots + \mu_t w_t \in \operatorname{Ker} f \subseteq \operatorname{Ker} f^{i+1}.$$

Now observe that we must have $\mu_1 w_1 + \cdots + \mu_t w_t = 0_V$, for otherwise, since

$$\{v_1, \ldots, v_s, w_1, \ldots, w_t\}$$

is a basis of $\operatorname{Ker} f^{i+1}$, there would exist scalars α_i and β_i, not all zero, such that

$$\sum_{i=1}^{t} \mu_i w_i = \sum_{i=1}^{s} \alpha_i v_i + \sum_{i=1}^{t} \beta_i w_i$$

whence we would have a dependence relation between the basis elements of the subspace $\operatorname{Ker} f^{i+1}$ and clearly this is not possible. Since $w_1, \ldots, w_t$ are linearly independent, it now follows from the equality $\mu_1 w_1 + \cdots + \mu_t w_t = 0_V$ that each $\mu_i = 0$. Consequently S is linearly independent. $\quad\square$

We now introduce a special type of upper triangular matrix.

Definition

By an **elementary Jordan matrix** associated with $\lambda \in F$ we shall mean either the 1×1 matrix $[\lambda]$ or a square matrix of the form

$$\begin{bmatrix} \lambda & 1 & 0 & \cdots & 0 & 0 \\ 0 & \lambda & 1 & \cdots & 0 & 0 \\ 0 & 0 & \lambda & \cdots & 0 & 0 \\ \vdots & \vdots & \vdots & \ddots & \vdots & \vdots \\ 0 & 0 & 0 & \cdots & \lambda & 1 \\ 0 & 0 & 0 & \cdots & 0 & \lambda \end{bmatrix}$$

in which all the diagonal entries are λ, all the entries immediately above the diagonal entries are 1, and all other entries are 0. By a **Jordan block matrix** associated with the eigenvalue $\lambda \in F$ we shall mean a matrix of the form

$$\begin{bmatrix} J_1 & & & \\ & J_2 & & \\ & & \ddots & \\ & & & J_k \end{bmatrix}$$

where each J_i is an elementary Jordan matrix associated with λ and all other entries are 0.

Theorem 5.4

Let V be a non-zero finite-dimensional vector space over a field F and let $f : V \rightarrow V$ be linear and nilpotent of index k. Then there is a basis of V with respect to which the matrix of f is a Jordan block matrix associated with the eigenvalue 0.

Proof

For $i = 0, \ldots, k$ let $W_i = \operatorname{Ker} f^i$. Since f is nilpotent of index k we have, by Theorem 5.2, the chain

$$\{0_V\} = W_0 \subset W_1 \subset W_2 \subset \cdots \subset W_{k-1} \subset W_k = V.$$

Choose a basis B_1 of $W_1 = \operatorname{Ker} f$ and extend this by $T_2 \subseteq W_2 \backslash W_1$ to a basis $B_2 = B_1 \cup T_2$ of W_2. Next, extend this by $T_3 \subseteq W_3 \backslash W_2$ to a basis $B_3 = B_2 \cup T_3$ of W_3, and so on. Then $B_k = B_1 \cup T_2 \cup \cdots \cup T_k$ is a basis of V.

Now let us work backwards, replacing each $T_i = B_i \backslash B_{i-1}$ as we go, except for T_k. To be more specific, let $T_k = \{x_1, \ldots, x_\alpha\}$. By Theorem 5.3 we have that $\{f(x_1), \ldots, f(x_\alpha)\}$ is a linearly independent subset of W_{k-1}. Moreover, it is disjoint from W_{k-2} since $f(x_i) \in W_{k-2} = \operatorname{Ker} f^{k-2}$ gives the contradiction $x_i \in \operatorname{Ker} f^{k-1} = W_{k-1}$. Consider therefore the set

$$B_{k-2} \cup \{f(x_1), \ldots, f(x_\alpha)\}.$$

This is linearly independent in W_{k-1} and so can be extended to a basis of W_{k-1}, say

$$B_{k-2} \cup \{f(x_1), \ldots, f(x_\alpha)\} \cup \{y_1, \ldots, y_\beta\}$$

where each $y_i \in W_{k-1} \backslash W_{k-2}$. In this way we have replaced T_{k-1} in the basis B_k by

$$T^\star_{k-1} = \{f(x_1), \ldots, f(x_\alpha)\} \cup \{y_1, \ldots, y_\beta\}.$$

Repeating the argument with the role of T_k assumed by $T^\star_{k-1}$, we can construct a basis of W_{k-2} of the form

$$B_{k-3} \cup \{f^2(x_1), \ldots, f^2(x_\alpha)\} \cup \{f(y_1), \ldots, f(y_\beta)\} \cup \{z_1, \ldots, z_\gamma\}$$

where each $z_i \in W_{k-2} \backslash W_{k-3}$. In this way we have replaced T_{k-2}.

Continuing in this way, and using the fact that a basis for $W_0 = \{0_V\}$ is $\emptyset$, we see that we can replace the basis B_k of V by the basis described in the following array:

$$
\begin{array}{llllll}
T_k : & x_1, & \ldots, & x_\alpha, & & \\
T_{k-1} \rightsquigarrow & f(x_1), & \ldots, & f(x_\alpha), & y_1, & \ldots, & y_\beta, \\
T_{k-2} \rightsquigarrow & f^2(x_1), & \ldots, & f^2(x_\alpha), & f(y_1), & \ldots, & f(y_\beta), z_1, \ldots, z_\gamma, \\
\vdots & & & & & \\
B_1 \rightsquigarrow & f^{k-1}(x_1), & \ldots, & f^{k-1}(x_\alpha), f^{k-2}(y_1), & \ldots, & f^{k-2}(y_\beta), & \ldots \ldots, q_1, \ldots, q_\omega.
\end{array}
$$

Note that in this array the elements in the bottom row form a basis for $W_1 = \operatorname{Ker} f$, those in the bottom two rows form a basis for $W_2 = \operatorname{Ker} f^2$, and so on. Also, the array is such that every element is mapped by f to the element lying immediately below it, the elements of the bottom row being mapped to 0_V.

We now order this basis of V by taking the first column starting at the bottom, then the second column, starting at the bottom, and so on. Then, as the reader can easily verify (using a larger sheet of paper!), the matrix of f relative to this ordered basis is a Jordan block matrix associated with the eigenvalue 0.   □

The above process is best illustrated by an example.

Example 5.2

Consider the linear mapping $f : \mathbb{R}^4 \to \mathbb{R}^4$ given by
$$f(a,b,c,d) = (0,a,d,0).$$
We have $f^2 = 0$ and so f is nilpotent of index $k = 2$. Now
$$W_1 = \operatorname{Ker} f = \{(0,b,c,0) \; ; \; b,c \in \mathbb{R}\};$$
$$W_2 = \operatorname{Ker} f^2 = \mathbb{R}^4.$$
A basis for W_1 is $B_1 = \{(0,1,0,0),(0,0,1,0)\}$ which we can extend to a basis
$$B_2 = B_1 \cup T_2 = \{(0,1,0,0),(0,0,1,0)\} \cup \{(1,0,0,0),(0,0,0,1)\}$$
of $W_2 = \mathbb{R}^4$. Now the image of T_2 under f is
$$\{(0,1,0,0),(0,0,1,0)\}$$
which is independent in W_1. In fact, this is the basis B_1, and so the array in the theorem becomes
$$(1,0,0,0), \quad (0,0,0,1),$$
$$(0,1,0,0), \quad (0,0,1,0).$$
Arranging the columns bottom-up and from left to right, we obtain the ordered basis
$$B = \{(0,1,0,0),(1,0,0,0),(0,0,1,0),(0,0,0,1)\}.$$
In order to compute the matrix of f relative to the ordered basis B, we observe that the transition matrix from B to the standard basis is
$$P = \begin{bmatrix} 0 & 1 & 0 & 0 \\ 1 & 0 & 0 & 0 \\ 0 & 0 & 1 & 0 \\ 0 & 0 & 0 & 1 \end{bmatrix}.$$
Now since $P^{-1} = P$ and the matrix of f relative to the standard basis is
$$A = \begin{bmatrix} 0 & 0 & 0 & 0 \\ 1 & 0 & 0 & 0 \\ 0 & 0 & 0 & 1 \\ 0 & 0 & 0 & 0 \end{bmatrix},$$

it follows that the matrix of f relative to B is the Jordan block matrix

$$P^{-1}AP = \begin{bmatrix} 0 & 1 \\ 0 & 0 \\ & & 0 & 1 \\ & & 0 & 0 \end{bmatrix}.$$

In practice, we rarely have to carry out the above computation. To see why, let us take a closer look at the proof of Theorem 5.5.

Relative to the ordered basis B constructed from the table we see that there are $\alpha \geqslant 1$ elementary Jordan matrices of size $k \times k$, then $\beta \geqslant 0$ of size $(k-1) \times (k-1)$, and so on. The number of elementary Jordan matrices involved is therefore

$$\alpha + \beta + \gamma + \cdots + \omega.$$

But, again from the table, this is precisely the number of elements in the bottom row. But the bottom row is a basis of $W_1 = \mathrm{Ker}\, f$. Consequently we see that *the number of elementary Jordan matrices involved is* $\dim \mathrm{Ker}\, f$.

Returning to Example 5.2, we see that $\mathrm{Ker}\, f$ has dimension 2, so there are only two elementary Jordan matrices involved. Since at least one of these has to be of size $k \times k = 2 \times 2$, the only possibility for the Jordan block matrix is

$$\begin{bmatrix} 0 & 1 \\ 0 & 0 \\ & & 0 & 1 \\ & & 0 & 0 \end{bmatrix}.$$

Our objective now is to extend the scope of Theorem 5.4 by removing the restriction that f be nilpotent. For this purpose, let us return to the Primary Decomposition Theorem. With the notation used there, let us assume that all the eigenvalues of f lie in the ground field F. Then we observe, by Theorem 3.2 and its Corollary 2, that *the minimum polynomial of f_i is* $(X - \lambda_i)^{e_i}$ and consequently *the mapping $f_i - \lambda_i \mathrm{id}_{V_i}$ is nilpotent of index e_i on the d_i-dimensional subspace V_i.*

Theorem 5.5

[Jordan Form]　*Let V be a non-zero finite-dimensional vector space over a field F and let $f : V \to V$ be linear. If $\lambda_1, \ldots, \lambda_k$ are the distinct eigenvalues of f and if each λ_i belongs to F then there is an ordered basis of V with respect to which the matrix of f is a block diagonal matrix*

$$\begin{bmatrix} J_1 \\ & J_2 \\ & & \ddots \\ & & & J_k \end{bmatrix}$$

in which J_i is a Jordan block matrix associated with λ_i.

Proof

With the usual notation, by Theorem 5.4 there is a basis of $V_i = \text{Ker}(f - \lambda_i \text{id}_{V_i})^{e_i}$ with respect to which the matrix of $f_i - \lambda_i \text{id}_{V_i}$ is a Jordan block matrix with 0 down the diagonal (since the only eigenvalue of a nilpotent mapping is 0). It follows that the matrix J_i of f_i is a Jordan block matrix with λ_i down the diagonal. $\square$

Definition

A matrix of the form described in Theorem 5.5 is called a **Jordan matrix** of f.

Strictly speaking, a Jordan matrix of f is not unique since the order in which the Jordan blocks J_i appear down the diagonal is not specified. However, the number of such blocks, the size of each block, and the number of elementary Jordan matrices that appear in each block, are uniquely determined by f. So, if we agree that in each Jordan block the elementary Jordan matrices are arranged down the diagonal in decreasing order of size, and the Jordan blocks themselves are arranged in increasing order of the magnitude of the eigenvalues, we have a form that we can refer to as 'the' Jordan matrix of f.

Definition

If $A \in \text{Mat}_{n \times n} F$ is such that all the eigenvalues of A belong to F then by the **Jordan normal form** of A we shall mean the Jordan matrix of any linear mapping represented by A relative to some ordered basis.

At this stage it is useful to retain the following summary of the above:

- If the characteristic and minimum polynomials of f (or of any square matrix A representing f) are $c = \prod_{i=1}^{k}(X - \lambda_i)^{d_i}$ and $m = \prod_{i=1}^{k}(X - \lambda_i)^{e_i}$ then, in the Jordan matrix of f (or Jordan normal form of A), the eigenvalue λ_i appears precisely d_i times down the diagonal, the number of elementary Jordan matrices associated with λ_i is $\dim \text{Ker}(f_i - \lambda_i \text{id}_{V_i})$, which is the geometric multiplicity of the eigenvalue λ_i, and at least one of each of these elementary Jordan matrices is of maximum size $e_i \times e_i$.

Example 5.3

Let $f : \mathbb{R}^7 \to \mathbb{R}^7$ be linear with characteristic and minimum polynomials

$$c_f = (X - 1)^3(X - 2)^4, \quad m_f = (X - 1)^2(X - 2)^3.$$

In any Jordan matrix that represents f the eigenvalue 1 appears three times down the diagonal with at least one associated elementary Jordan matrix of size 2×2; and the eigenvalue 2 appears four times down the diagonal with at least one associated elementary Jordan matrix of size 3×3. Up to the order of the blocks, there is therefore only one possibility for the Jordan normal form, namely

$$\begin{bmatrix} 1 & 1 & & & & \\ & 1 & & & & \\ & & 1 & & & \\ & & & 2 & 1 & \\ & & & & 2 & 1 \\ & & & & & 2 \\ & & & & & & 2 \end{bmatrix}.$$

Example 5.4

Let us modify the previous example slightly. Suppose that c_f is as before but that now

$$m_f = (X - 1)^2 (X - 2)^2.$$

In this case the eigenvalue 2 appears four times in the diagonal with at least one associated elementary Jordan matrix of size 2×2. The possibilities for the Jordan normal form are then

$$\begin{bmatrix} 1 & 1 & & & & \\ & 1 & & & & \\ & & 1 & & & \\ & & & 2 & 1 & \\ & & & & 2 & \\ & & & & & 2 & 1 \\ & & & & & & 2 \end{bmatrix}, \quad \begin{bmatrix} 1 & 1 & & & & \\ & 1 & & & & \\ & & 1 & & & \\ & & & 2 & 1 & \\ & & & & 2 & \\ & & & & & 2 & \\ & & & & & & 2 \end{bmatrix}.$$

Example 5.5

If $f : \mathbb{R}^5 \to \mathbb{R}^5$ has characteristic polynomial

$$c_f = (X - 2)^2 (X - 3)^3$$

then the possible Jordan normal forms, which are obtained by considering all six possible minimum polynomials, are $\begin{bmatrix} A & \\ & B \end{bmatrix}$ where A is one of

$$\begin{bmatrix} 2 & 1 \\ & 2 \end{bmatrix}, \quad \begin{bmatrix} 2 & \\ & 2 \end{bmatrix}$$

and B is one of

$$\begin{bmatrix} 3 & 1 & \\ & 3 & 1 \\ & & 3 \end{bmatrix}, \quad \begin{bmatrix} 3 & 1 & \\ & 3 & \\ & & 3 \end{bmatrix}, \quad \begin{bmatrix} 3 & & \\ & 3 & \\ & & 3 \end{bmatrix}.$$

Example 5.6

For the matrix

$$A = \begin{bmatrix} 1 & 3 & -2 \\ 0 & 7 & -4 \\ 0 & 9 & -5 \end{bmatrix}$$

we have $c_A = (X - 1)^3$ and $m_A = (X - 1)^2$. The Jordan normal form is then

$$\begin{bmatrix} 1 & 1 & \\ & 1 & \\ & & 1 \end{bmatrix}.$$

Example 5.7

Consider the matrix

$$A = \begin{bmatrix} 2 & 1 & 1 & 1 & 0 \\ 0 & 2 & 0 & 0 & 0 \\ 0 & 0 & 2 & 1 & 0 \\ 0 & 0 & 0 & 1 & 1 \\ 0 & -1 & -1 & -1 & 0 \end{bmatrix}.$$

As can readily be seen, the characteristic polynomial of A is

$$c_A = (X - 1)^3 (X - 2)^2.$$

The general eigenvector associated with the eigenvalue 1 is $[0, 0, x, -x, 0]$ so the corresponding eigenspace has dimension 1. Likewise, the general eigenvector associated with the eigenvalue 2 is $[x, y, -y, 0, 0]$ so the corresponding eigenspace has dimension 2. Thus, in the Jordan normal form, the eigenvalue 1 appears three times down the diagonal with only one associated elementary Jordan matrix; and the eigenvalue 2 appears twice down the diagonal with two associated elementary Jordan matrices. Consequently the Jordan normal form of A is

$$\begin{bmatrix} 1 & 1 & & & \\ & 1 & 1 & & \\ & & 1 & & \\ & & & 2 & \\ & & & & 2 \end{bmatrix}.$$

EXERCISES

5.3 Determine the Jordan normal form of the matrix

$$\begin{bmatrix} -13 & 8 & 1 & 2 \\ -22 & 13 & 0 & 3 \\ 8 & -5 & 0 & -1 \\ -22 & 13 & 5 & 5 \end{bmatrix}.$$

5.4 Determine the Jordan normal form of the matrix
$$\begin{bmatrix} 5 & -1 & -3 & 2 & -5 \\ 0 & 2 & 0 & 0 & 0 \\ 1 & 0 & 1 & 1 & -2 \\ 0 & -1 & 0 & 3 & 1 \\ 1 & -1 & -1 & 1 & 1 \end{bmatrix}.$$

5.5 Find the Jordan normal form of the differentiation map on the vector space of real polynomials of degree at most 3.

5.6 Let V be the vector space of functions $q : \mathbb{R}^2 \to \mathbb{R}$ of the form
$$q(x,y) = ax^2 + bxy + cy^2 + dx + ey + f.$$

Let $\varphi : V \to V$ be the linear mapping given by
$$\varphi(q) = \frac{\partial}{\partial x} \int q(x,y)\, dy.$$

Determine the matrix M that represents φ relative to the ordered basis B of V given by
$$B = \{x^2, xy, y^2, x, y, 1\}.$$

Compute

(1) the characteristic and minimum polynomials of M;

(2) the Jordan normal form J of M.

Definition

By a **Jordan basis** for $f : V \to V$ we shall mean an ordered basis of V with respect to which the matrix of f is a Jordan matrix.

To obtain a method of computing a Jordan basis, consider first the $t \times t$ elementary Jordan matrix
$$\begin{bmatrix} \lambda & 1 & & & & \\ & \lambda & 1 & & & \\ & & \lambda & 1 & & \\ & & & \ddots & \ddots & \\ & & & & \lambda & 1 \\ & & & & & \lambda \end{bmatrix}.$$

A corresponding basis $\{v_1, \ldots, v_t\}$ will be such that
$$f(v_1) = \lambda v_1;$$
$$f(v_2) = \lambda v_2 + v_1;$$
$$f(v_3) = \lambda v_3 + v_2;$$
$$\vdots$$
$$f(v_{t-1}) = \lambda v_{t-1} + v_{t-2};$$
$$f(v_t) = \lambda v_t + v_{t-1}.$$

Thus, for every $t \times t$ elementary Jordan matrix associated with λ we require $v_1, \ldots, v_t$ to be linearly independent with

(1) $v_1 \in \mathrm{Im}\,(f - \lambda\,\mathrm{id}) \cap \mathrm{Ker}\,(f - \lambda\,\mathrm{id})$;
(2) $(i = 2, \ldots, t)$ $(f - \lambda\,\mathrm{id})(v_i) = v_{i-1}$.

The solution in the general case is then obtained by applying the above procedure to each elementary Jordan matrix and pasting together the resulting ordered bases.

Example 5.8

Let $f : \mathbb{R}^3 \to \mathbb{R}^3$ be given by

$$f(x, y, z) = (x + y, -x + 3y, -x + y + 2z).$$

Relative to the standard ordered basis of $\mathbb{R}^3$, the matrix of f is

$$A = \begin{bmatrix} 1 & 1 & 0 \\ -1 & 3 & 0 \\ -1 & 1 & 2 \end{bmatrix}.$$

We have $c_A = (X - 2)^2$ and $m_A = (X - 2)^2$. The Jordan normal form is then

$$J = \begin{bmatrix} 2 & 1 & \\ & 2 & \\ & & 2 \end{bmatrix}.$$

Now we have

$$(f - 2\,\mathrm{id})(x, y, z) = (-x + y, -x + y, -x + y)$$

and we begin by choosing $v_1 \in \mathrm{Im}\,(f - 2\,\mathrm{id}) \cap \mathrm{Ker}\,(f - 2\,\mathrm{id})$. Clearly, $v_1 = (1, 1, 1)$ will do. Next we have to find v_2, independent of v_1, such that $(f - 2\,\mathrm{id})(v_2) = v_1$. Clearly, $v_2 = (1, 2, 1)$ will do. To complete the basis, we now have to choose $v_3 \in \mathrm{Ker}\,(f - 2\,\mathrm{id})$ with $\{v_1, v_2, v_3\}$ independent. Clearly, $v_3 = (1, 1, 0)$ will do.

Thus a Jordan basis is

$$B = \{(1, 1, 1), (1, 2, 1), (1, 1, 0)\}.$$

To determine an invertible matrix P such that $P^{-1}AP = J$, it suffices to observe that the transition matrix from B to the standard basis is

$$P = \begin{bmatrix} 1 & 1 & 1 \\ 1 & 2 & 1 \\ 1 & 1 & 0 \end{bmatrix}.$$

Then a simple calculation reveals that $P^{-1}AP = J$.

Example 5.9

An ordered basis for $\mathbb{R}_4[X]$ is $\{1, X, X^2, X^3, X^4\}$. Relative to this the differentiation map D is represented by the matrix

$$A = \begin{bmatrix} 0 & 1 & 0 & 0 & 0 \\ 0 & 0 & 2 & 0 & 0 \\ 0 & 0 & 0 & 3 & 0 \\ 0 & 0 & 0 & 0 & 4 \\ 0 & 0 & 0 & 0 & 0 \end{bmatrix}.$$

The characteristic polynomial of A is X^5, the only eigenvalue is 0, and the eigenspace of 0 is of dimension 1 with basis $\{1\}$. So the Jordan normal form of A is

$$J = \begin{bmatrix} 0 & 1 & 0 & 0 & 0 \\ 0 & 0 & 1 & 0 & 0 \\ 0 & 0 & 0 & 1 & 0 \\ 0 & 0 & 0 & 0 & 1 \\ 0 & 0 & 0 & 0 & 0 \end{bmatrix}.$$

A Jordan basis is $\{p_1, p_2, p_3, p_4, p_5\}$ where

$$Dp_1 = 0, \ Dp_2 = p_1, \ Dp_3 = p_2, \ Dp_4 = p_3, \ Dp_5 = p_4.$$

We may choose $p_1 = 1$ and $p_2 = X$, then $p_3 = \frac{1}{2}X^2$, $p_4 = \frac{1}{6}X^3$, $p_5 = \frac{1}{24}X^4$. A Jordan basis is therefore $\{24, 24X, 12X^2, 4X^3, X^4\}$.

EXERCISES

5.7 Determine a Jordan basis for the linear mappings represented by the matrices in

(1) Exercise 5.4;

(2) Exercise 5.5.

5.8 Referring to Exercise 5.6, determine

(1) a Jordan basis for φ;

(2) an invertible matrix P such that $P^{-1}MP = J$ where J is the Jordan normal form of M.

5.9 Suppose that $f : \mathbb{R}^5 \to \mathbb{R}^5$ is represented with respect to the ordered basis

$$\{(1,0,0,0,0), (1,1,0,0,0), (1,1,1,0,0), (1,1,1,1,0), (1,1,1,1,1)\}$$

by the matrix

$$A = \begin{bmatrix} 1 & 8 & 6 & 4 & 1 \\ 0 & 1 & 0 & 0 & 0 \\ 0 & 1 & 2 & 1 & 0 \\ 0 & -1 & -1 & 0 & 1 \\ 0 & -5 & -4 & -3 & -2 \end{bmatrix}.$$

Find a basis of $\mathbb{R}^5$ with respect to which the matrix of f is in Jordan normal form.

An interesting consequence of the Jordan normal form is the following.

Theorem 5.6

Every square matrix A over $\mathbb{C}$ is similar to its transpose.

Proof

Since all the eigenvalues of A belong to $\mathbb{C}$ it clearly suffices to establish the result for the Jordan form of A. Because of the structure of this, it is enough to establish the result for an elementary Jordan matrix of the form

$$J = \begin{bmatrix} \lambda & 1 & & & \\ & \lambda & 1 & & \\ & & \ddots & \ddots & \\ & & & \lambda & 1 \\ & & & & \lambda \end{bmatrix}.$$

Now if

$$B = \{v_1, \ldots, v_k\}$$

is an associated Jordan basis, define

$$(i = 1, \ldots, k) \qquad w_i = v_{k-i+1}$$

and consider the ordered basis

$$B^\star = \{w_1, \ldots, w_k\} = \{v_k, \ldots, v_1\}.$$

It is readily seen that the matrix relative to this basis is the transpose of J. Consequently we see that J is similar to its transpose. $\square$

By way of an application, we shall now illustrate the usefulness of the Jordan normal form in solving systems of linear differential equations. Here it is not our intention to become heavily involved with the theory. A little by way of explanation together with an illustrative example is all we have in mind.

By a **system of linear differential equations with constant coefficients** we shall mean a system of equations of the form

$$x_1' = a_{11}x_1 + a_{12}x_2 + \cdots + a_{1n}x_n$$

$$x_2' = a_{21}x_1 + a_{22}x_2 + \cdots + a_{2n}x_n$$

$$\vdots$$

$$x_n' = a_{n1}x_1 + a_{n2}x_2 + \cdots + a_{nn}x_n$$

where $x_1, \ldots, x_n$ are real differentiable functions of t, x_i' denotes the derivative of x_i, and $a_{ij} \in \mathbb{R}$ for all i, j.

These equations can be written in the matrix form

$$(1) \qquad \mathbf{X}' = A\mathbf{X}$$

where $\mathbf{X} = [x_1 \cdots x_n]^t$, and $A = [a_{ij}]_{n \times n}$.

Suppose that A can be reduced to Jordan normal form J_A, and let P be an invertible matrix such that $P^{-1}AP = J_A$. Writing $\mathbf{Y} = P^{-1}\mathbf{X}$, we have

$$(2) \qquad (P\mathbf{Y})' = \mathbf{X}' = A\mathbf{X} = AP\mathbf{Y}$$

and so

$$(3) \qquad \mathbf{Y}' = P^{-1}\mathbf{X}' = P^{-1}AP\mathbf{Y} = J_A\mathbf{Y}.$$

Now the form of J_A means that (3) is a system that is considerably easier to solve for $\mathbf{Y}$. Then, by (2), $P\mathbf{Y}$ is a solution of (1).

Example 5.10

Consider the system

$$x_1' = x_1 + x_2$$

$$x_2' = -x_1 + 3x_2$$

$$x_3' = -x_1 + 4x_2 - x_3$$

i.e. $\mathbf{X}' = A\mathbf{X}$ where

$$\mathbf{X} = \begin{bmatrix} x_1 \\ x_2 \\ x_3 \end{bmatrix}, \qquad A = \begin{bmatrix} 1 & 1 & 0 \\ -1 & 3 & 0 \\ -1 & 4 & -1 \end{bmatrix}.$$

We have

$$c_A = (X + 1)(X - 2)^2 = m_A$$

and so the Jordan form of A is

$$J_A = \begin{bmatrix} -1 & & \\ & 2 & 1 \\ & & 2 \end{bmatrix}.$$

We now determine an invertible matrix P such that $P^{-1}AP = J_A$. For this, we determine a Jordan basis. For a change, let us do so with matrices rather than mappings. Clearly, we have to find independent column vectors $\mathbf{p}_1, \mathbf{p}_2, \mathbf{p}_3$ such that

$$(A + I_3)\mathbf{p}_1 = \mathbf{0},$$
$$(A - 2I_3)\mathbf{p}_2 = \mathbf{0},$$
$$(A - 2I_3)\mathbf{p}_3 = \mathbf{p}_2.$$

Suitable vectors are, for example,

$$\mathbf{p}_1 = \begin{bmatrix} 0 \\ 0 \\ 1 \end{bmatrix}, \quad \mathbf{p}_2 = \begin{bmatrix} 1 \\ 1 \\ 1 \end{bmatrix}, \quad \mathbf{p}_3 = \begin{bmatrix} -1 \\ 0 \\ 0 \end{bmatrix}.$$

Thus we may take

$$P = \begin{bmatrix} 0 & 1 & -1 \\ 0 & 1 & 0 \\ 1 & 1 & 0 \end{bmatrix}.$$

[The reader should check that $P^{-1}AP = J_A$ or, what is equivalent and much easier, that $AP = PJ_A$.]

With $\mathbf{Y} = P^{-1}\mathbf{X}$ we now solve the system $\mathbf{Y}' = J_A\mathbf{Y}$, i.e.

$$y_1' = -y_1,$$
$$y_2' = 2y_2 + y_3,$$
$$y_3' = 2y_3.$$

The first and third of these equations give

$$y_1 = \alpha_1 e^{-t}, \quad y_3 = \alpha_3 e^{2t};$$

and the second equation becomes

$$y_2' = 2y_2 + \alpha_3 e^{2t}.$$

It follows that

$$y_2 = \alpha_3 t e^{2t} + \alpha_2 e^{2t}.$$

Consequently we see that

$$\mathbf{Y} = \begin{bmatrix} \alpha_1 e^{-t} \\ \alpha_2 e^{2t} + \alpha_3 t e^{2t} \\ \alpha_3 e^{2t} \end{bmatrix}.$$

A solution of the original system of equations is then given by

$$\mathbf{X} = P\mathbf{Y} = \begin{bmatrix} \alpha_2 e^{2t} + \alpha_3(t-1)e^{2t} \\ \alpha_2 e^{2t} + \alpha_3 t e^{2t} \\ \alpha_1 e^{-t} + \alpha_2 e^{2t} + \alpha_3 t e^{2t} \end{bmatrix}.$$

EXERCISES

5.10 Find the Jordan normal form J of the matrix

$$A = \begin{bmatrix} 0 & 1 & 0 & -1 \\ -2 & 3 & 0 & -1 \\ -2 & 1 & 2 & -1 \\ 2 & -1 & 0 & 3 \end{bmatrix}.$$

Find also a Jordan basis and an invertible matrix P such that $P^{-1}AP = J$.

Hence solve the system of differential equations

$$\begin{bmatrix} x_1' \\ x_2' \\ x_3' \\ x_4' \end{bmatrix} = \begin{bmatrix} 0 & 1 & 0 & -1 \\ -2 & 3 & 0 & -1 \\ -2 & 1 & 2 & -1 \\ 2 & -1 & 0 & 3 \end{bmatrix} \begin{bmatrix} x_1 \\ x_2 \\ x_3 \\ x_4 \end{bmatrix}.$$

5.11 Solve the system of differential equations

$$x_1' = x_1 + 3x_2 - 2x_3$$
$$x_2' = 7x_2 - 4x_3$$
$$x_3' = 9x_2 - 5x_3.$$

5.12 Show how the differential equation

$$x''' - 2x'' - 4x' + 8x = 0$$

can be written in the matrix form $\mathbf{X}' = A\mathbf{X}$ where

$$A = \begin{bmatrix} 0 & 1 & 0 \\ 0 & 0 & 1 \\ -8 & 4 & 2 \end{bmatrix}.$$

[*Hint.* Write $x_1 = x$, $x_2 = x'$, $x_3 = x''$.]

By using the method of the Jordan form, solve the equation given the initial conditions

$$x(0) = 0, \quad x'(0) = 0, \quad x''(0) = 16.$$

6
Rational and Classical Forms

Although in general the minimum polynomial of a linear mapping $f : V \to V$ can be expressed as a product of powers of irreducible polynomials over the ground field F of V, say $m_f = p_1^{e_1} p_2^{e_2} \cdots p_k^{e_k}$, the irreducible polynomials p_i need not be linear. Put another way, the eigenvalues of f need not belong to the ground field F. It is therefore natural to seek a canonical matrix representation for f in the general case, which will reduce to the Jordan representation when all the eigenvalues of f do belong to F. In order to develop the machinery to deal with this, we first consider the following notion.

Suppose that W is a subspace of the F-vector space V. Then in particular W is a (normal) subgroup of the additive (abelian) group of V and so we can form the quotient group V/W. The elements of this are the cosets

$$x + W = \{x + w \; ; \; w \in W\},$$

and the group operation on V/W is given by

$$(x + W) + (y + W) = (x + y) + W.$$

Clearly, under this operation the natural surjection $\natural_W : V \to V/W$ given by $\natural_W(x) = x + W$ is a group morphism. Can we define a multiplication by scalars in such a way that V/W becomes a vector space over F and the natural surjection $\natural_W$ is linear? Indeed we can, and in only one way. In fact, for $\natural_W$ to be linear it is necessary that we have the identity

$$(\forall x \in V)(\forall \lambda \in F) \qquad \lambda \natural_W(x) = \natural_W(\lambda x),$$

and so multiplication by scalars must be given by

$$\lambda(x + W) = \lambda x + W.$$

With respect to this operation of multiplication by scalars it is readily verified that the additive abelian group V/W becomes a vector space over F. We call this the **quotient space of V by W** and denote it also by V/W.

As we shall now show, if V is of finite dimension then so also is every quotient space of V.

Theorem 6.1

Let V be a finite-dimensional vector space and let W be a subspace of V. Then the quotient space V/W is also finite-dimensional. Moreover, if $\{v_1, \ldots, v_m\}$ is a basis of W and $\{x_1 + W, \ldots, x_k + W\}$ is a basis of V/W then $\{v_1, \ldots, v_m, x_1, \ldots, x_k\}$ is a basis of V.

Proof

Suppose that $I = \{x_1 + W, \ldots, x_p + W\}$ is any linearly independent subset of V/W. Then the set $\{x_1, \ldots, x_p\}$ of coset representatives is a linearly independent subset of V. To see this, suppose that $\sum_{i=1}^{p} \lambda_i x_i = 0_V$. Then, using the linearity of $\natural_W$, we have

$$0_{V/W} = \natural_W(0_V) = \natural_W\left(\sum_{i=1}^{p} \lambda_i x_i\right) = \sum_{i=1}^{p} \lambda_i \natural_W(x_i) = \sum_{i=1}^{p} \lambda_1(x_i + W)$$

and so each λ_i is 0. Consequently, $p \leqslant \dim V$. Since $|I| = p$ it follows that every linearly independent subset of V/W has at most $\dim V$ elements. Hence V/W is of finite dimension.

Suppose now that $\{v_1, \ldots, v_m\}$ is a basis of W and that $\{x_1 + W, \ldots, x_k + W\}$ is a basis of V/W. Consider the set $B = \{v_1, \ldots, v_m, x_1, \ldots, x_k\}$. Applying $\natural_W$ to any linear combination of elements of B we see as above that B is linearly independent. Now for every $x \in V$ we have $x + W \in V/W$ and so there exist scalars λ_i such that

$$x + W = \sum_{i=1}^{k} \lambda_i(x_i + W) = \left(\sum_{i=1}^{k} \lambda_i x_i\right) + W$$

and hence $x - \sum_{i=1}^{k} \lambda_i x_i \in W$. Then there exist μ_j such that $x - \sum_{i=1}^{k} \lambda_i x_i = \sum_{j=1}^{m} \mu_j v_j$. Consequently x is a linear combination of the elements of B. It follows that the linearly independent set B is also a spanning set and therefore is a basis of V. $\square$

Corollary 1

$\dim V = \dim W + \dim V/W$.

Proof

$\dim V = |B| = m + k = \dim W + \dim V/W$. $\square$

Corollary 2

If $V = W \oplus Z$ then $Z \simeq V/W$.

Proof

We have $\dim V = \dim W + \dim Z$ and so, by Corollary 1, $\dim Z = \dim V/W$ whence it follows that $Z \simeq V/W$. $\square$

EXERCISES

6.1 Let V and W be vector spaces over a field F and let $f : V \to W$ be linear. If Z is a subspace of V prove that the assignment

$$x + Z \mapsto f(x)$$

defines a linear mapping from V/Z to W if and only if $Z \subseteq \text{Ker } f$.

6.2 If $f : V \to W$ is linear prove that $V/\text{Ker } f \simeq \text{Im} f$.

We shall be particularly interested in the quotient space V/W when W is a subspace that is f-invariant for a given linear mapping $f : V \to V$. In this situation we have the following result.

Theorem 6.2

Let V be a finite-dimensional vector space and let $f : V \to V$ be linear. If W is an f-invariant subspace of V then the prescription

$$f^+(x + W) = f(x) + W$$

defines a linear mapping $f^+ : V/W \to V/W$, the minimum polynomial of which divides the minimum polynomial of f.

Proof

Observe that if $x + W = y + W$ then $x - y \in W$ and so, since W is f-invariant, we have $f(x) - f(y) = f(x - y) \in W$ from which we obtain $f(x) + W = f(y) + W$. Thus the above prescription does indeed define a mapping f^+ from V/W to itself.

That f^+ is linear follows from the fact that, for all $x, y \in V$ and all scalars λ,

$$
\begin{aligned}
f^+[(x + W) + (y + W)] &= f^+[(x + y) + W] \\
&= f(x + y) + W \\
&= [f(x) + f(y)] + W \\
&= [f(x) + W] + [f(y) + W] \\
&= f^+(x + W) + f^+(y + W); \\
f^+[\lambda(x + W)] &= f^+[\lambda x + W] \\
&= f(\lambda x) + W \\
&= \lambda f(x) + W \\
&= \lambda[f(x) + W] \\
&= \lambda f^+(x + W).
\end{aligned}
$$

To show that the minimum polynomial of f^+ divides that of f, we now show by induction that

$$(\forall n \geqslant 0) \qquad (f^+)^n = (f^n)^+.$$

For the anchor point $n = 1$ this is clear. As for the inductive step, suppose that $(f^+)^n = (f^n)^+$. Then, for every $x \in V$,

$$
\begin{aligned}
(f^+)^{n+1}(x + W) &= f^+[(f^+)^n(x + W)] \\
&= f^+[(f^n)^+(x + W)] \\
&= f^+[f^n(x) + W] \\
&= f[f^n(x)] + W \\
&= f^{n+1}(x) + W \\
&= (f^{n+1})^+(x + W),
\end{aligned}
$$

and consequently $(f^+)^{n+1} = (f^{n+1})^+$.

It follows from this that for every polynomial $p = \sum_{i=0}^{m} a_i X^i$ we have

$$p(f^+) = \sum_{i=0}^{m} a_i (f^+)^i = \sum_{i=0}^{m} a_i (f^i)^+ = [p(f)]^+.$$

Taking in particular $p = m_f$ we obtain $m_f(f^+) = 0$ whence $m_{f^+} | m_f$. $\quad\square$

Definition

We call $f^+ : V/W \to V/W$ the linear mapping that is **induced** by f on the quotient space V/W.

EXERCISE

6.3 For the linear mapping $f : \mathbb{R}^3 \to \mathbb{R}^3$ given by $f(x, y, z) = (x, x, x)$ describe the induced mapping $f^+ : \mathbb{R}^3/\mathrm{Ker}\, f \to \mathbb{R}^3/\mathrm{Ker}\, f$.

We shall now focus on a particular type of invariant subspace. Suppose that V is a finite-dimensional vector space and that $f : V \to V$ is linear. It is readily seen that the intersection of any family of f-invariant subspaces of V is also an f-invariant subspace of V. It follows immediately from this that for every subset X of V there is a smallest f-invariant subspace that contains X, namely the intersection of all the f-invariant subspaces that contain X. We shall denote this by Z_X^f. In the case where $X = \{x\}$ we shall write Z_x^f, or simply Z_x when f is clearly understood. The subspace Z_x can be characterised as follows.

Theorem 6.3

Let V be a finite-dimensional vector space over a field F and let $f : V \to V$ be linear. Then, for every $x \in V$,

$$Z_x = \{p(f)(x) \; ; \; p \in F[X]\}.$$

Proof

It is readily seen that the set $W = \{p(f)(x) \; ; \; p \in F[X]\}$ is a subspace of V that contains x. Since f commutes with every $p(f)$, this subspace is f-invariant.

Suppose now that T is an f-invariant subspace that contains x. Then clearly T contains $f^k(x)$ for all k, and consequently contains $p(f)(x)$ for every polynomial $p \in F[X]$. Thus T contains W. Hence W is the smallest f-invariant subspace that contains x and so coincides with Z_x. $\square$

Example 6.1

Let $f : \mathbb{R}^3 \to \mathbb{R}^3$ be given by

$$f(a,b,c) = (-b + c, \, a + c, \, 2c).$$

Consider the element $(1,0,0)$. We have $f(1,0,0) = (0,1,0)$ and $f^2(1,0,0) = f(0,1,0) = -(1,0,0)$, from which it is readily seen that

$$Z_{(1,0,0)} = \{(x,y,0) \; ; \; x,y \in \mathbb{R}\}.$$

EXERCISE

6.4 Let $f : \mathbb{R}^3 \to \mathbb{R}^3$ be given by

$$f(a,b,c) = (0, a, b).$$

Determine Z_x where $x = (1,-1,3)$.

Our immediate objective is to discover a basis for the subspace Z_x. For this purpose, consider the sequence

$$x, \, f(x), \, f^2(x), \, \dots, \, f^r(x), \, \dots$$

of elements of Z_x. Clearly, there exists a smallest positive integer k such that $f^k(x)$ is a linear combination of the elements that precede it in this list, say

$$f^k(x) = \lambda_0 x + \lambda_1 f(x) + \cdots + \lambda_{k-1}f^{k-1}(x),$$

and $\{x, f(x), \dots, f^{k-1}(x)\}$ is then a linearly independent subset of Z_x.

Writing $a_i = -\lambda_i$ for $i = 0, \dots, k-1$ we deduce that the polynomial

$$m_x = a_0 + a_1 X + \cdots + a_{k-1}X^{k-1} + X^k$$

is the monic polynomial of least degree such that $m_x(f)(x) = 0_V$.

Definition

We call m_x the f-**annihilator** of x.

Example 6.2

Referring to Example 6.1, let $x = (1, 0, 0)$. Then, as we have seen, $f^2(x) = -x$. It follows that the f-annihilator of x is the polynomial $m_x = X^2 + 1$.

EXERCISES

6.5 In Exercise 6.4, determine the f-annihilator of $x = (1, -1, 3)$.

6.6 Let $f : \mathbb{R}^3 \to \mathbb{R}^3$ be the linear mapping that is represented, relative to the standard ordered basis of $\mathbb{R}^3$, by the matrix

$$\begin{bmatrix} 1 & 0 & 1 \\ 0 & 1 & 0 \\ -1 & 0 & -1 \end{bmatrix}.$$

Determine the f-annihilator of $x = (1, 1, 1)$.

With the above notation we now have the following result.

Theorem 6.4

Let V be a finite-dimensional vector space and let $f : V \to V$ be linear. If $x \in V$ has f-annihilator

$$m_x = a_0 + a_1 X + \cdots + a_{k-1} X^{k-1} + X^k$$

then the set

$$B_x = \{x, f(x), \ldots, f^{k-1}(x)\}$$

is a basis of Z_x and therefore $\dim Z_x = \deg m_x$. Moreover, if $f_x : Z_x \to Z_x$ is the induced linear mapping on the f-invariant subspace Z_x then the matrix of f_x relative to the ordered basis B_x is

$$C_{m_x} = \begin{bmatrix} 0 & 0 & 0 & \cdots & 0 & -a_0 \\ 1 & 0 & 0 & \cdots & 0 & -a_1 \\ 0 & 1 & 0 & \cdots & 0 & -a_2 \\ \vdots & \vdots & \vdots & & \vdots & \vdots \\ 0 & 0 & 0 & \cdots & 1 & -a_{k-1} \end{bmatrix}.$$

Finally, the minimum polynomial of f_x is m_x.

Proof

Clearly, B_x is linearly independent and $f^k(x) \in \text{Span } B_x$. We prove by induction that in fact $f^n(x) \in \text{Span } B_x$ for every n. This being clear for $n = 1, \ldots, k$ suppose that $n > k$ and that $f^{n-1}(x) \in \text{Span } B_x$. Then $f^{n-1}(x)$ is a linear combination of $x, f(x), \ldots, f^{k-1}(x)$ and so $f^n(x)$ is a linear combination of $f(x), f^2(x), \ldots, f^k(x)$, whence it follows that $f^n(x) \in \text{Span } B_x$. It is immediate from this observation that

$p(f)(x) \in$ Span B_x for every polynomial p. Thus $Z_x \subseteq$ Span B_x whence we have equality since the reverse inclusion is trivial. Consequently, B_x is a basis of Z_x.

Now since

$$f_x(x) = f(x)$$

$$f_x[f(x)] = f^2(x)$$

$$\vdots$$

$$f_x[f^{k-2}(x)] = f^{k-1}(x)$$

$$f_x[f^{k-1}(x)] = f^k(x) = -a_0 x - a_1 f(x) - \cdots - a_{k-1} f^{k-1}(x)$$

it is clear that the matrix of f_x relative to the basis B_x is the above matrix C_{m_x}.

Finally, suppose that the minimum polynomial of f_x is

$$m_{f_x} = b_0 + b_1 X + \cdots + b_{r-1} X^{r-1} + X^r.$$

Then

$$0_V = m_{f_x}(f)(x) = b_0 x + b_1 f(x) + \cdots + b_{r-1} f^{r-1}(x) + f^r(x)$$

from which we see that $f^r(x)$ is a linear combination of $x, f(x), \ldots, f^{r-1}(x)$ and therefore $k \leqslant r$. But m_{f_x} is the zero map on Z_x, whence so is $m_x(f_x)$. Consequently we have $m_{f_x} | m_x$ and so $r \leqslant k$. Thus $r = k$ and it follows that $m_{f_x} = m_x$. $\square$

Definition

If V is a finite-dimensional vector space and $f : V \rightarrow V$ is linear then a subspace W of V is said to be f-**cyclic** if it is f-invariant and has a basis of the form $\{x, f(x), \ldots, f^m(x)\}$. Such a basis is called a **cyclic basis**, and x is called a **cyclic vector** for W. In particular, Theorem 6.4 shows that $x \in V$ is a cyclic vector for the subspace Z_x with cyclic basis B_x. The subspace Z_x is called the f-**cyclic subspace spanned by** $\{x\}$. The matrix C_{m_x} of Theorem 6.4 is called the **companion matrix** of the f-annihilator m_x.

Our first main objective can now be revealed. Bringing the above results together, we shall show that if $f : V \rightarrow V$ has minimum polynomial of the form p^k where p is irreducible then V can be expressed as a direct sum of f-cyclic subspaces. The main consequence of this is that f then has a block diagonal representation by companion matrices. One more observation will lead us to this goal.

Theorem 6.5

Let W be an f-invariant subspace of V. Then for every $x \in V$ both the f-annihilator of f and the f^+-annihilator of $x + W$ divide the minimum polynomial of f.

Proof

By Theorem 6.4, the f-annihilator of x is the minimum polynomial of f_x, the mapping induced on Z_x by f, which clearly divides the minimum polynomial of f.

As for the f^+-annihilator of $x+W$, this likewise divides the minimum polynomial of f^+ which, by Theorem 6.2, divides that of f. □

Theorem 6.6

[Cyclic Decomposition]   *Let V be a non-zero vector space of finite dimension over a field F and let $f : V \to V$ be linear with minimum polynomial $m_f = p^t$ where p is irreducible over F. Then there are f-cyclic vectors $x_1, \ldots, x_k$ and positive integers $n_1, \ldots, n_k$ with each $n_i \leqslant t$ such that*

(1) $V = \displaystyle\bigoplus_{i=1}^{k} Z_{x_i}$;

(2) *the f-annihilator of x_i is p^{n_i}.*

Proof

The proof is by induction on $\dim V$. The result is trivial when $\dim V = 1$. Suppose then that the result holds for all vector spaces of dimension less than n (where $n > 1$) and let V be of dimension n.

Since $m_f = p^t$ there is a non-zero $x_1 \in V$ with $p^{t-1}(f)(x_1) \neq 0_V$. The f-annihilator of x_1 is then $m_{x_1} = p^t$. Let $W = Z_{x_1}$ and let $f^+ : V/W \to V/W$ be the induced linear mapping. By Theorem 6.2, the minimum polynomial of f^+ divides $m_f = p^t$ and so the inductive hypothesis applies to f^+ and V/W. Thus there exist f^+-cyclic subspaces $Z_{y_2+W}, \ldots, Z_{y_k+W}$ of V/W such that

$$V/W = \bigoplus_{i=2}^{k} Z_{y_i+W}$$

and, for $2 \leqslant i \leqslant k$, the f^+-annihilator of $y_i + W$ is p^{n_i} for some $n_i \leqslant t$.

We now observe that there exists $x_i \in y_i + W$ such that the f-annihilator of x_i is p^{n_i}. In fact, since the f^+-annihilator of $y_i + W$ is p^{n_i}, we have $p(f)^{n_i}(y_i) \in W = Z_{x_1}$. Thus there is a polynomial h such that

$$p(f)^{n_i}(y_i) = h(f)(x_1).$$

It follows from this that

$$0_V = p(f)^t(y_i) = p(f)^{t-n_i} h(f)(x_1).$$

But p^t is the f-annihilator of x_1, so $p^t | p^{t-n_i} h$ and hence $p^{n_i} | h$. Consequently $h = p^{n_i} q$ for some polynomial q. Now define $x_i = y_i - q(f)(x_1)$. Then

$$y_i - x_i = q(f)(x_1) \in W$$

and so $x_i \in y_i + W$ whence $x_i + W = y_i + W$. The f^+-annihilator of $y_i + W$ therefore divides the f-annihilator of x_i. But

$$p(f)^{n_i}(x_i) = p(f)^{n_i}[y_i - q(f)(x_1)] = p(f)^{n_i}(y_i) - h(f)(x_1) = 0_V.$$

Thus we see that the f-annihilator of x_i is p^{n_i}.

Suppose now that $\deg p = d$. Then $\deg p^{n_i} = dn_i$. Now since p^{n_i} is both the f-annihilator of x_i and the f^+-annihilator of $x_i + W$, it follows by Theorem 6.4 that

$$A_i = \{x_i, f(x_i), \ldots, f^{dn_i-1}(x_i)\}$$

is a basis for Z_{x_i}, and that

$$B_i = \{x_i + W, f^+(x_i + W), \ldots, (f^+)^{dn_i-1}(x_i + W)\}$$

is a basis for Z_{x_i+W}. But since

$$V/W = \bigoplus_{i=2}^{k} Z_{y_i+W} = \bigoplus_{i=2}^{k} Z_{x_i+W}$$

it follows that $\bigcup\limits_{i=2}^{k} B_i$ is a basis of V/W. Then, by Theorem 6.1, $\bigcup\limits_{i=1}^{k} A_i$ is a basis of V.

Consequently we have that $V = \bigoplus\limits_{i=1}^{k} Z_{x_i}$ and the induction is complete. $\square$

Corollary 1

With the above notation, relative to the basis $\bigcup\limits_{i=1}^{k} A_i$ the matrix of f is of the form

$$\bigoplus_{i=1}^{k} C_i = \begin{bmatrix} C_1 & & & \\ & C_2 & & \\ & & \ddots & \\ & & & C_k \end{bmatrix}$$

where C_i is the companion matrix of $m_{x_i} = p^{n_i}$. $\square$

Corollary 2

$\dim V = (n_1 + \cdots + n_k) \deg p.$ $\square$

Without loss of generality, we may assume that the f-cyclic vectors $x_1, \ldots, x_k$ of Theorem 6.6 are arranged such that the corresponding integers n_i satisfy

$$t = n_1 \geqslant n_2 \geqslant \cdots \geqslant n_k \geqslant 1.$$

With this convention we have the following result.

Theorem 6.7

The integers $n_1, \ldots, n_k$ are uniquely determined by f.

Proof

From the above we have, for every i,

$$\dim Z_{x_i} = \deg m_{x_i} = \deg p^{n_i} = dn_i.$$

Observe now that for every j the image of Z_{x_i} under $p(f)^j$ is the f-cyclic subspace $Z_{p(f)^j(x_i)}$. Since the f-annihilator of x_i is p^{n_i}, of degree dn_i, we see that

$$\dim Z_{p(f)^j(x_i)} = \begin{cases} 0 & \text{if } j \geqslant n_i; \\ d(n_i - j) & \text{if } j < n_i. \end{cases}$$

Now every $x \in V$ can be written uniquely in the form

$$x = v_1 + \cdots + v_k \qquad (v_i \in Z_{x_i})$$

and so every element of $\text{Im}\, p(f)^j$ can be written uniquely in the form

$$p(f)^j(x) = p(f)^j(v_1) + \cdots + p(f)^j(v_k).$$

Thus, if r is the integer such that $n_1, \ldots, n_r > j$ and $n_{r+1} \leqslant j$ then we see that

$$\text{Im}\, p(f)^j = \bigoplus_{i=1}^{r} Z_{p(f)^j(x_i)}$$

and consequently

$$\dim \text{Im}\, p(f)^j = d \sum_{i=1}^{r} (n_i - j) = d \sum_{n_i > j} (n_i - j).$$

It follows from this that

$$\dim \text{Im}\, p(f)^{j-1} - \dim \text{Im}\, p(f)^j = d\Big(\sum_{n_i > j-1} (n_i - j + 1) - \sum_{n_i > j} (n_i - j)\Big)$$

$$= d \sum_{n_i \geqslant j} (n_i - j + 1 - n_i + j)$$

$$= d \sum_{n_i \geqslant j} 1$$

$$= d \times (\text{the number of } n_i \geqslant j).$$

Now the dimensions on the left of this are determined by f so the above expression gives, for each j, the number of n_i that are greater than or equal to j. This determines the sequence

$$t = n_1 \geqslant n_2 \geqslant \cdots \geqslant n_k \geqslant 1$$

completely. □

Definition

When the minimum polynomial of f is of the form p^t where p is irreducible then, relative to the uniqely determined chain of integers $t = n_1 \geqslant n_2 \geqslant \cdots \geqslant n_k \geqslant 1$ as described above, the polynomials $p^t = p^{n_1}, p^{n_2}, \ldots, p^{n_k}$ are called the **elementary divisors** of f.

It should be noted that the first elementary divisor in the sequence is the minimum polynomial of f.

We can now apply the above results to the general situation where the characteristic and minimum polynomials of a linear mapping $f : V \to V$ are

$$c_f = p_1^{d_1} p_2^{d_2} \cdots p_k^{d_k}, \qquad m_f = p_1^{e_1} p_2^{e_2} \cdots p_k^{e_k}$$

where $p_1, \ldots, p_k$ are distinct irreducible polynomials.

We know by the Primary Decomposition Theorem that there is an ordered basis of V with respect to which the matrix of f is a block diagonal matrix

$$\begin{bmatrix} A_1 & & & \\ & A_2 & & \\ & & \ddots & \\ & & & A_k \end{bmatrix}$$

in which each A_i is the matrix (of size $d_i \deg p_i \times d_i \deg p_i$) that represents the induced mapping f_i on $V_i = \mathrm{Ker}\, p_i(f)^{e_i}$. Now the minimum polynomial of f_i is $p_i^{e_i}$ and so, by the Cyclic Decomposition Theorem, there is a basis of V_i with respect to which A_i is the block diagonal matrix

$$\begin{bmatrix} C_{i1} & & & \\ & C_{i2} & & \\ & & \ddots & \\ & & & C_{it} \end{bmatrix}$$

in which the C_{ij} are the companion matrices associated with the elementary divisors of f_i. By the previous discussion, this block diagonal form, in which each block A_i is itself a block diagonal of companion matrices, is unique (to within the order of the A_i). It is called the **rational canonical matrix** of f.

It is important to note that in the sequence of elementary divisors there may be repetitions, for some of the n_i can be equal. The result of this is that some companion matrices can appear more than once in the rational form.

Example 6.3

Suppose that $f : \mathbb{R}^4 \to \mathbb{R}^4$ is linear with minimum polynomial

$$m_f = X^2 + 1.$$

Then the characteristic polynomial must be $c_f = (X^2 + 1)^2$. Since $X^2 + 1$ is irreducible over $\mathbb{R}$ it follows by Corollary 2 of Theorem 6.6 that $4 = (n_1 + \cdots + n_k)2$. Now since the first elementary divisor is the minimum polynomial we must have $n_1 = 1$. Since we must also have each $n_i \geqslant 1$, it follows that the only possibility is $k = 2$ and

$n_1 = n_2 = 1$. The rational canonical matrix of f is therefore

$$C_{X^2+1} \oplus C_{X^2+1} = \begin{bmatrix} 0 & -1 & & \\ 1 & 0 & & \\ & & 0 & -1 \\ & & 1 & 0 \end{bmatrix}.$$

Example 6.4

Suppose now that $f : \mathbb{R}^6 \to \mathbb{R}^6$ has minimum polynomial

$$m_f = (X^2 + 1)(X - 2)^2.$$

The characteristic polynomial of f is then one of

$$c_1 = (X^2 + 1)^2(X - 2)^2, \qquad c_2 = (X^2 + 1)(X - 2)^4.$$

Suppose first that $c_f = c_1$. In this case we have $\mathbb{R}^6 = V_1 \oplus V_2$ with dim $V_1 = 4$ and dim $V_2 = 2$. The induced mapping f_1 on V_1 has minimum polynomial $m_1 = X^2 + 1$ and the induced mapping f_2 on V_2 has minimum polynomial $m_2 = (X - 2)^2$. The situation for V_1 is as in the previous Example. As for V_2, it follows by Corollary 2 of Theorem 6.6 that $2 = n_1 + \cdots + n_k$ whence necessarily $k = 1$ since $n_1 = \deg p_2 = 2$. Thus the only elementary divisor of f_2 is $(X - 2)^2$. Combining these observations, we see that in this case the rational canonical matrix of f is

$$C_{X^2+1} \oplus C_{X^2+1} \oplus C_{(X-2)^2}.$$

Suppose now that $c_f = c_2$. In this case we have $\mathbb{R}^6 = V_1 \oplus V_2$ with dim $V_1 = 2$ and dim $V_2 = 4$. Also, the induced mapping f_2 on V_2 has minimum polynomial $m_2 = (X-2)^2$. By Corollary 2 of Theorem 6.6 applied to f_2, we have $4 = n_1 + \cdots + n_k$ with $n_1 = 2$. There are therefore two possibilities, namely

$$k = 2 \text{ with } n_1 = n_2 = 2;$$

$$k = 3 \text{ with } n_1 = 2, n_2 = n_3 = 1.$$

The rational canonical matrix of f in this case is therefore one of the forms

$$C_{X^2+1} \oplus C_{(X-2)^2} \oplus C_{(X-2)^2};$$

$$C_{X^2+1} \oplus C_{(X-2)^2} \oplus C_{X-2} \oplus C_{X-2}.$$

Note from the above Example that a knowledge of both the characteristic and minimum polynomials is not in general enough to determine completely the rational form.

The reader will not fail to notice that the rational form is quite different from the Jordan form. To illustrate this more drammatically, let us take a matrix in Jordan form and find its rational form.

Example 6.5

Consider the matrix

$$A = \begin{bmatrix} 2 & 1 & 0 \\ 0 & 2 & 1 \\ 0 & 0 & 2 \end{bmatrix}.$$

We have $c_A = (X - 2)^3 = m_A$ and, by Corollary 2 of Theorem 6.6, $3 = n_1 + \cdots + n_k$ with $n_1 = 3$. Thus $k = 1$ and the rational form is

$$C_{(X-2)^3} = \begin{bmatrix} 0 & 0 & 8 \\ 1 & 0 & -12 \\ 0 & 1 & 6 \end{bmatrix}.$$

EXERCISES

6.7 Determine the rational canonical form of the matrix

$$A = \begin{bmatrix} a & 0 & 1 \\ 0 & a & 1 \\ -1 & 1 & a \end{bmatrix}.$$

6.8 Let $f : \mathbb{R}^3 \rightarrow \mathbb{R}^3$ be the linear mapping that is represented, relative to the standard ordered basis of $\mathbb{R}^3$, by the matrix

$$A = \begin{bmatrix} 1 & 2 & 2 \\ 2 & 1 & 2 \\ 2 & 2 & 1 \end{bmatrix}.$$

Determine the rational canonical matrix of f.

6.9 Let $f : \mathbb{R}^3 \rightarrow \mathbb{R}^3$ be the linear mapping that is represented, relative to the standard ordered basis of $\mathbb{R}^3$, by the matrix

$$A = \begin{bmatrix} 5 & -6 & -6 \\ -1 & 4 & 2 \\ 3 & -6 & -4 \end{bmatrix}.$$

Determine the rational canonical matrix of f.

6.10 Suppose that $f : \mathbb{R}^7 \rightarrow \mathbb{R}^7$ is linear with characteristic and minimum polynomials

$$c_f = (X - 1)^3 (X - 2)^4, \qquad m_f = (X - 1)^2 (X - 2)^3.$$

Determine the rational canonical matrix of f.

6.11 If V is a finite-dimensional vector space and $f : V \rightarrow V$ is linear, prove that V has an f-cyclic vector if and only if $c_f = m_f$.

The fact that the rational form is quite different from the Jordan form suggests that we are not quite finished, for what we seek is a general canonical form that will reduce to the Jordan form when all the eigenvalues belong to the ground field. We shall now proceed to obtain such a form by modifying the cyclic bases that were used to obtain the rational form. In so doing, we shall obtain a matrix representation that is constructed from the companion matrix of p_i rather than from that of $p_i^{n_i}$.

Theorem 6.8

Let x be a cyclic vector of V and let $f : V \to V$ be linear with minimum polynomial p^n where

$$p = \alpha_0 + \alpha_1 X + \cdots + \alpha_{k-1} X^{k-1} + X^k.$$

Then there is a basis of V with respect to which the matrix of f is the $kn \times kn$ matrix

$$\begin{bmatrix} C_p & M & & & \\ & C_p & M & & \\ & & \ddots & \ddots & \\ & & & C_p & M \\ & & & & C_p \end{bmatrix}$$

in which C_p is the companion matrix of p and M is the $k \times k$ matrix

$$\begin{bmatrix} & & 1 \\ & & \\ & & \end{bmatrix}.$$

Proof

Consider the $n \times k$ array

$$\begin{array}{cccc} f^{k-1}(x) & \cdots & f(x) & x \\ p(f)[f^{k-1}(x)] & \cdots & p(f)[f(x)] & p(f)(x) \\ \vdots & & \vdots & \vdots \\ p(f)^{n-1}[f^{k-1}(x)] & \cdots & p(f)^{n-1}[f(x)] & p(f)^{n-1}(x) \end{array}$$

We first show that this forms a basis of V. For this purpose, it suffices to show that the elements in the array are linearly independent. Suppose, by way of obtaining a contradiction, that this were not so. Then some non-trivial linear combination of these elements would be 0_V and consequently there would exist a polynomial h such that $h(f)(x) = 0_V$ with $\deg h < kn = \deg p^n$. Since x is cyclic, this contradicts the hypothesis that p^n is the minimum polynomial of f. Hence the above array constitutes a basis for V.

We order this basis in a row-by-row manner, as we normally read.

Observe now that f maps each element in the array to its predecessor in the same row, except for those at the beginning of a row. For these elements we have

$$f[f^{k-1}(x)] = f^k(x) = -\alpha_{k-1}f^{k-1}(x) - \cdots - \alpha_0 x + p(f)(x)$$

and similarly, writing $p(f)$ as h for convenience,

$$f[h^{n-m}(f^{k-1})(x)] = h^{n-m}[f^k(x)]$$
$$= h^{n-m}[-\alpha_{k-1}f^{k-1}(x) - \cdots - \alpha_0 x + p(f)(x)]$$
$$= -\alpha_{k-1}h^{n-m}[f^{k-1}(x)] - \cdots - \alpha_0 h^{n-m}(x) + h^{n-m+1}(x).$$

It is now a simple matter to verify that the matrix of f relative to the above ordered basis is of the form described. $\quad\square$

Definition

The block matrix described in Theorem 6.8 is called the **classical p-matrix** associated with the companion matrix C_p.

Moving now to general considerations, consider a linear mapping $f : V \to V$. With the usual notation, the induced mapping f_i on V_i has minimum polynomial $p_i^{e_i}$. If we now apply Theorem 6.8 to the cyclic subspaces that appear in the Cyclic Decomposition Theorem for f_i, we see that in the rational canonical matrix of f_i we can replace each diagonal block of companion matrices C_{ij} associated with the elementary divisors $p_i^{n_i}$ by the classical p_i-matrix associated with the companion matrix of p_i. The matrix that arises in this way is called the **classical canonical matrix** of f.

Example 6.6

Let $f : \mathbb{R}^6 \to \mathbb{R}^6$ be linear and such that

$$c_f = m_f = (X^2 - X + 1)^2(X + 1)^2.$$

By the Primary Decomposition Theorem we have that the minimum polynomial of the induced mapping f_1 is $(X^2 - X + 1)^2$ whereas that of the induced mapping f_2 is $(X + 1)^2$. Using Corollary 2 of Theorem 6.6 we see that the only elementary divisor of f_1 is $(X^2 - X + 1)^2$, and the only elementary divisor of f_2 is $(X + 1)^2$. Consequently the rational canonical matrix of f is

$$C_{(X^2-X+1)^2} \oplus C_{(X+1)^2} = \begin{bmatrix} 0 & 0 & 0 & -1 & & \\ 1 & 0 & 0 & 2 & & \\ 0 & 1 & 0 & -3 & & \\ 0 & 0 & 1 & 2 & & \\ & & & & 0 & -1 \\ & & & & 1 & -2 \end{bmatrix}.$$

The classical canonical matrix of f can be obtained from this as follows. With $p_1 = X^2 - X + 1$ we replace the companion matrix associated with p_1^2 by the classical p_1-matrix associated with the companion matrix of p_1. We do a similar replacement concerning $p_2 = X + 1$. The result is

$$
\begin{bmatrix}
C_{X^2-X+1} & M_{2\times 2} & & \\
& C_{X^2-X+1} & & \\
& & C_{X+1} & M_{1\times 1} \\
& & & C_{X+1}
\end{bmatrix}
=
\begin{bmatrix}
0 & -1 & 0 & 1 & & & \\
1 & 1 & 0 & 0 & & & \\
& & 0 & -1 & & & \\
& & 1 & 1 & & & \\
& & & & -1 & 1 & \\
& & & & & -1 &
\end{bmatrix}.
$$

Finally, let us take note of the particular case of Theorem 6.8 that occurs when $k = 1$. In this situation we have $p = X - \alpha_0$, and $f - \alpha_0 \mathrm{id}_V$ is nilpotent of index n. Consequently C_p reduces to the 1×1 matrix $[\alpha_0]$ and the classical p-matrix associated with C_p reduces to the $n \times n$ elementary Jordan matrix associated with the eigenvalue α_0. Thus we see that when the eigenvalues all belong to the ground field the classical form reduces to the Jordan form.

Example 6.7

Let $f : \mathbb{R}^7 \to \mathbb{R}^7$ be linear and such that

$$c_f = (X - 1)^3(X - 2)^4, \qquad m_f = (X - 1)^2(X - 2)^3.$$

The rational canonical matrix is

$$
C_{(X-1)^2} \oplus C_{X-1} \oplus C_{(X-2)^3} \oplus C_{X-2} =
\begin{bmatrix}
0 & 1 & & & & & \\
1 & 2 & & & & & \\
& & 1 & & & & \\
& & & & & & 8 \\
& & & 1 & & & -12 \\
& & & & 1 & & 6 \\
& & & & & & 2
\end{bmatrix}.
$$

Here all the eigenvalues belong to the ground field, so the classical canonical matrix coincides with the Jordan normal matrix. The reader can verify that it is

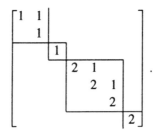

EXERCISES

6.12 Prove that if $g \in F[X]$ is monic of degree n and if C_g is the companion matrix of g then

$$\det(XI_n - C_g) = g.$$

[*Hint*. Use induction on n. Let

$$g = a_0 + a_1 X + \cdots + a_{n-1}X^{n-1} + X^n$$

and let $h \in F[X]$ be given by

$$h = a_1 + a_2 X + \cdots + a_{n-1}X^{n-2} + X^{n-1}.$$

Show that $\det(XI_n - C_g) = Xh + a_0$.]

6.13 For each of the following matrices over $\mathbb{R}$ determine

(a) the characteristic polynomial;

(b) the minimum polynomial;

(c) the elementary divisors;

(d) the rational canonical form;

(e) the classical canonical form.

$$\begin{bmatrix} 1 & 1 & 3 \\ 5 & 2 & 6 \\ -2 & -1 & -3 \end{bmatrix} ; \qquad \begin{bmatrix} 2 & 2 & 1 \\ 2 & 2 & 1 \\ 2 & 2 & 1 \end{bmatrix} ; \qquad \begin{bmatrix} 0 & 0 & 1 & 1 \\ 0 & 0 & 1 & 1 \\ 1 & 1 & 0 & 0 \\ 1 & 1 & 0 & 0 \end{bmatrix}.$$

6.14 Let V be a real vector space of dimension 10 and let $f : V \to V$ be linear and such that

$$c_f = (X^2 - X + 1)^3(X + 1)^4, \qquad m_f = (X^2 - X + 1)^2(X + 1)^2.$$

Determine the rational and classical canonical matrices of f.

6.15 Repeat the previous exercise assuming now that V is a complex vector space.

7
Dual Spaces

If V and W are vector spaces over the same field F, consider the set $\mathrm{Lin}\,(V, W)$ that consists of all the linear mappings $f : V \to W$. Given $f, g \in \mathrm{Lin}\,(V, W)$, define the mapping $f + g : V \to W$ by the prescription

$$(\forall x \in V) \qquad (f + g)(x) = f(x) + g(x).$$

Then it is readily seen that $f + g \in \mathrm{Lin}\,(V, W)$. Also, given $f \in \mathrm{Lin}\,(V, W)$ and $\lambda \in F$, define the mapping $\lambda f : V \to W$ by the prescription

$$(\forall x \in V) \qquad (\lambda f)(x) = \lambda\, f(x).$$

Then it is readily seen that $\lambda f \in \mathrm{Lin}\,(V, W)$. Moreover, under these operations of addition and multiplication by scalars, $\mathrm{Lin}\,(V, W)$ becomes a vector space over F.

A particular case of this is of especial importance, namely that in which for W we take the ground field F (regarded as a vector space over itself). It is on this vector space $\mathrm{Lin}\,(V, F)$ that we shall now focus our attention.

Definition

We shall call $\mathrm{Lin}\,(V, F)$ the **dual space** of V and denote it by V^d. The elements of V^d, i.e. the linear mappings from V to the ground field F, are called **linear forms** (or **linear functionals**) on V.

Example 7.1

The i-th projection $p_i : \mathbb{R}^n \to \mathbb{R}$ given by $p_i(x_1, \ldots, x_n) = x_i$ is a linear form on $\mathbb{R}^n$ and so is an element of $(\mathbb{R}^n)^d$.

Example 7.2

If V is the vector space of $n \times n$ matrices over a field F then the mapping $\mathrm{tr} : V \to F$ that sends each $A \in V$ to its trace $\mathrm{tr}\, A = \sum_{i=1}^{n} a_{ii}$ is a linear form on V and so is an element of V^d.

Example 7.3

The mapping $I : \mathbb{R}[X] \to \mathbb{R}$ given by $I(p) = \int_0^1 p$ is a linear form on $\mathbb{R}[X]$ and so is an element of $\mathbb{R}[X]^d$.

In what follows we shall often denote a typical element of V^d by x^d. The reader should therefore take careful note of the fact that this notation will be used to denote a linear mapping from V to the ground field F.

We begin by showing that if V is of finite dimension then so is its dual space V^d. This we do by constructing a basis for V^d from a basis of V.

Theorem 7.1

Let $\{v_1, \ldots, v_n\}$ be a basis of V and for $i = 1, \ldots, n$ let $v_i^d : V \to F$ be the linear mapping such that

$$v_i^d(v_j) = \delta_{ij} = \begin{cases} 1_F & \text{if } j = i; \\ 0_F & \text{if } j \neq i. \end{cases}$$

Then $\{v_1^d, \ldots, v_n^d\}$ is a basis of V^d.

Proof

It is clear that each $v_i^d \in V^d$. Suppose that $\sum_{i=1}^{n} \lambda_i v_i^d = 0$ in V^d. Then for $j = 1, \ldots, n$ we have

$$0_F = \left(\sum_{i=1}^{n} \lambda_i v_i^d \right)(v_j) = \sum_{i=1}^{n} \lambda_i v_i^d(v_j) = \sum_{i=1}^{n} \lambda_i \delta_{ij} = \lambda_j,$$

from which we see that $\{v_1^d, \ldots, v_n^d\}$ is linearly independent.

If now $x = \sum_{j=1}^{n} x_j v_j \in V$ then we have

$$(1) \qquad v_i^d(x) = \sum_{j=1}^{n} x_j v_i^d(v_j) = \sum_{j=1}^{n} x_j \delta_{ij} = x_i$$

and hence, for every $f \in V^d$,

$$\left(\sum_{i=1}^{n} f(v_i) v_i^d \right)(x) = \sum_{i=1}^{n} f(v_i) v_i^d(x) = \sum_{i=1}^{n} f(v_i) x_i = f\left(\sum_{i=1}^{n} x_i v_i \right) = f(x).$$

Thus we see that

$$(2) \qquad (\forall f \in V^d) \qquad f = \sum_{i=1}^{n} f(v_i) v_i^d,$$

which shows that $\{v_1^d, \ldots, v_n^d\}$ also spans V^d, whence it is a basis. $\quad\square$

Corollary

If V is a finite-dimensional vector space then $\dim V^d = \dim V$. $\quad\square$

Note from the equalities (1) and (2) in the above proof that we have

$$(\forall x \in V) \qquad x = \sum_{i=1}^{n} v_i^d(x) v_i;$$

$$(\forall x^d \in V^d) \qquad x^d = \sum_{i=1}^{n} x^d(v_i) v_i^d.$$

Definition

If $\{v_1, \ldots, v_n\}$ is a basis of V then by the corresponding **dual basis** we shall mean the basis $\{v_1^d, \ldots, v_n^d\}$ described in Theorem 6.1.

Because of (1) above, the mappings $v_1^d, \ldots, v_n^d$ are often called the **coordinate forms** associated with $v_1, \ldots, v_n$.

Example 7.4

Consider the basis $\{v_1, v_2\}$ of $\mathbb{R}^2$ where $v_1 = (1, 2)$ and $v_2 = (2, 3)$. Let $\{v_1^d, v_2^d\}$ be the corresponding dual basis. Then we have

$$1 = v_1^d(v_1) = v_1^d(1, 2) = v_1^d(1, 0) + 2v_1^d(0, 1);$$
$$0 = v_1^d(v_2) = v_1^d(2, 3) = 2v_1^d(1, 0) + 3v_1^d(0, 1).$$

These equations give $v_1^d(1, 0) = -3$ and $v_1^d(0, 1) = 2$ and hence v_1^d is given by

$$v_1^d(x, y) = -3x + 2y.$$

Similarly, we have

$$v_2^d(x, y) = 2x - y.$$

Example 7.5

Consider the standard basis $\{e_1, \ldots, e_n\}$ of $\mathbb{R}^n$. By definition, we have $e_i^d(e_j) = \delta_{ij}$ and so

$$e_i^d(x_1, \ldots, x_n) = e_i^d\left(\sum_{j=1}^{n} x_j e_j\right) = \sum_{j=1}^{n} x_j e_i^d(e_j) = x_i$$

whence the corresponding dual basis is the set of projections $\{p_1, \ldots, p_n\}$.

Example 7.6

Let $t_1, \ldots, t_{n+1}$ be distinct real numbers and for each i let $\zeta_{t_i} : \mathbb{R}_n[X] \to \mathbb{R}$ be the substitution mapping, given by $\zeta_{t_i}(p) = p(t_i)$. Then $B = \{\zeta_{t_1}, \ldots, \zeta_{t_{n+1}}\}$ is a basis for $\mathbb{R}_n[X]^d$. To see this, observe that since $\mathbb{R}_n[X]^d$ has the same dimension as $\mathbb{R}_n[X]$

(namely $n+1$), it suffices to show that B is linearly independent. Now if $\sum_{i=1}^{n+1} \lambda_i \zeta_{t_i} = 0$ then we have

$$0 = \left(\sum_{i=1}^{n+1} \lambda_i \zeta_{t_i}\right)(1) = \lambda_1 + \lambda_2 + \cdots + \lambda_{n+1}$$

$$0 = \left(\sum_{i=1}^{n+1} \lambda_i \zeta_{t_i}\right)(X) = \lambda_1 t_1 + \lambda_2 t_2 + \cdots + \lambda_{n+1} t_{n+1}$$

$$\vdots$$

$$0 = \left(\sum_{i=1}^{n+1} \lambda_i \zeta_{t_i}\right)(X^n) = \lambda_1 t_1^n + \lambda_2 t_2^n + \cdots + \lambda_{n+1} t_{n+1}^n.$$

This system of equations can be written in the matrix form $M\mathbf{x} = \mathbf{0}$ where $\mathbf{x} = [\lambda_1 \ \ldots \ \lambda_{n+1}]^t$ and M is the **Vandermonde matrix**

$$M = \begin{bmatrix} 1 & 1 & \cdots & 1 \\ t_1 & t_2 & \cdots & t_{n+1} \\ \vdots & \vdots & & \vdots \\ t_1^n & t_2^n & \cdots & t_{n+1}^n \end{bmatrix}.$$

By induction, it can be shown that

$$\det M = \prod_{j<i}(t_i - t_j).$$

Since $t_1, \ldots, t_{n+1}$ are distinct, it follows that $\det M \neq 0$. The only solution of $M\mathbf{x} = \mathbf{0}$ is therefore $\mathbf{x} = \mathbf{0}$ whence $\lambda_1 = \cdots = \lambda_{n+1} = 0$ as required.

To determine a basis of $\mathbb{R}_n[X]$ of which B is the dual, let such a basis be

$$A = \{p_1, \ldots, p_{n+1}\}.$$

Then we require $\zeta_{t_i}(p_j) = \delta_{ij}$ for all i, j. It is readily seen that the **Lagrange polynomials**

$$L_j = \prod_{i \neq j} \frac{X - t_i}{t_j - t_i}$$

are the successful candidates.

EXERCISES

7.1 If $B = \{E_{i,j} \ ; \ i = 1, \ldots, m$ and $j = 1, \ldots, n\}$ is the canonical basis of $\mathrm{Mat}_{m \times n} F$, determine the corresponding dual basis.

7.2 Let V be the vector space of $n \times n$ matrices over a field F. If $B \in V$ prove that $\varphi_B : V \to F$ given by $\varphi_B(A) = \mathrm{tr}\, B^t A$ is a linear form on V. Prove also that every $\varphi \in V^d$ is of this form for some $B \in V$.

Theorem 7.2

Let $(v_i)_n$, $(w_i)_n$ be ordered bases of V and let $(v_i^d)_n$, $(w_i^d)_n$ be the corresponding dual bases. If P is the transition matrix from $(v_i)_n$ to $(w_i)_n$ then the transition matrix from $(v_i^d)_n$ to $(w_i^d)_n$ is $(P^{-1})^t$.

Proof

Let the transition matrix from $(v_i^d)_n$ to $(w_i^d)_n$ be Q. Then we have, for each i,

$$w_i = \sum_{j=1}^n p_{ji} v_j, \qquad w_i^d = \sum_{j=1}^n q_{ji} v_j^d.$$

Consequently,

$$\delta_{ij} = w_i^d(w_j) = \sum_{k=1}^n q_{ki}\left(\sum_{t=1}^n p_{tj} v_k^d(v_t)\right)$$

$$= \sum_{k=1}^n q_{ki}\left(\sum_{t=1}^n p_{tj}\delta_{kt}\right)$$

$$= \sum_{k=1}^n q_{ki} p_{kj}$$

$$= [Q^t P]_{ij}.$$

Thus we see that $Q^t P = I_n$ and therefore $Q = (P^{-1})^t$. $\square$

Theorem 7.2 provides a very convenient way of exhibiting dual bases which we shall now describe.

Given a basis

$$B = \{(a_{11}, \ldots, a_{1n}), (a_{21}, \ldots, a_{2n}), \ldots, (a_{n1}, \ldots, a_{nn})\}$$

of $\mathbb{R}^n$, the transition matrix from B to the standard basis $(e_i)_n$ is the matrix P whose i-th column is $[a_{i1} \ \ldots \ a_{in}]^t$. By Theorem 6.2, if we associate with the i-th **row** of P^{-1} the mapping $[\pi_{i1}, \ldots, \pi_{in}] : \mathbb{R}^n \to \mathbb{R}$ defined by

$$[\pi_{i1}, \ldots, \pi_{in}](x_1, \ldots, x_n) = \pi_{i1} x_1 + \cdots + \pi_{in} x_n$$

then we see that

$$[\pi_{i1}, \ldots, \pi_{in}](a_{j1}, \ldots, a_{jn}) = [P^{-1} P]_{ij} = \delta_{ij}$$

and hence we can represent the dual basis of B by

$$B^d = \{[\pi_{11}, \ldots, \pi_{1n}], [\pi_{21}, \ldots, \pi_{2n}], \ldots, [\pi_{n1}, \ldots, \pi_{nn}]\}.$$

Example 7.7

Consider again Example 7.4 where we computed the basis that is dual to the basis $B = \{(1,2), (2,3)\}$ of $\mathbb{R}^2$. The transition matrix from B to the standard basis is

$$P = \begin{bmatrix} 1 & 2 \\ 2 & 3 \end{bmatrix}$$

and its inverse is

$$P^{-1} = \begin{bmatrix} -3 & 2 \\ 2 & -1 \end{bmatrix}.$$

The basis that is dual to B can therefore be described as

$$B^d = \{[-3, 2], [2, -1]\}$$

where the notation is defined such that

$$[-3, 2](x, y) = -3x + 2y, \qquad [2, -1](x, y) = 2x - y.$$

Example 7.8

Consider again Example 7.5 where we computed the basis that is dual to the standard basis $B = (e_i)_n$ of $\mathbb{R}^n$. Applying the above procedure, we see that $P = I_n = P^{-1}$ from which it follows that the dual basis consists of the projections $p_1, \ldots, p_n$.

EXERCISES

7.3 Determine the basis of $(\mathbb{R}^3)^d$ that is dual to the basis

$$\{(1, 0, -1), (-1, 1, 0), (0, 1, 1)\}.$$

7.4 Determine the basis of $(\mathbb{R}^4)^d$ that is dual to the basis

$$\{(4, 5, -2, 11), (3, 4, -2, 6), (2, 3, -1, 4), (1, 1, -1, 3)\}.$$

7.5 Let $A = \{x_1, x_2\}$ be a basis of a vector space V of dimension 2, and let $A^d = (\varphi_1, \varphi_2)$ be the corresponding dual basis of V^d. Find, in terms of φ_1 and φ_2, the basis of V^d that is dual to the basis

$$B = \{x_1 + 2x_2, 3x_1 + 4x_2\}$$

of V.

We now introduce the following notation. Given $x \in V$ and $y^d \in V^d$ we shall write

$$y^d(x) = \langle x, y^d \rangle.$$

With this notation, the following identities are immediate from the linearity of the mappings involved:

(α) $\langle x + y, z^d \rangle = \langle x, z^d \rangle + \langle y, z^d \rangle$;

(β) $\langle x, y^d + z^d \rangle = \langle x, y^d \rangle + \langle x, z^d \rangle$;

(γ) $\langle \lambda x, y^d \rangle = \lambda \langle x, y^d \rangle$;

(δ) $\langle x, \lambda y^d \rangle = \lambda \langle x, y^d \rangle$.

Now it is clear from (β) and (δ) that for every $x \in V$ the mapping $\hat{x} : V^d \to F$ defined by

$$\hat{x}(y^d) = \langle x, y^d \rangle$$

is a linear form on V^d, i.e. is an element of the dual space $\widehat{V} = (V^d)^d$ of V^d.

Moreover, by (α) and (γ) it is quickly verified that the mapping $\alpha_V : V \to \widehat{V}$ defined by $\alpha_V(x) = \hat{x}$ is also linear.

Definition

We call $\widehat{V}$ the **bidual** of V, and α_V the **canonical mapping** from V to $\widehat{V}$.

- Note that the various notations employed above can be summarised by the following identities:

$$\boxed{\hat{x}(y^d) = \langle x, y^d \rangle = y^d(x)}.$$

Theorem 7.3

If V is of finite dimension then the canonical mapping $\alpha_V : V \to \widehat{V}$ is an isomorphism.

Proof

Let $\{v_1, \ldots, v_n\}$ be a basis of V and consider $x = \sum_{j=1}^{n} x_j v_j \in \text{Ker } \alpha_V$. Then $\hat{x}$ is the zero of $\widehat{V}$ and so $\hat{x}(y^d) = 0$ for all $y^d \in V^d$. In particular, for $i = 1, \ldots, n$ we have

$$0 = \hat{x}(v_i^d) = \langle x, v_i^d \rangle = v_i^d(x) = \sum_{j=1}^{n} x_j v_i^d(v_j) = x_i$$

and consequently $x = 0_V$. Thus α_V is injective.

Now by the Corollary of Theorem 7.1 we have

$$\dim V = \dim V^d = \dim \widehat{V}.$$

It follows therefore that α_V is an isomorphism. $\square$

In the case where V is of finite dimension we shall agree to identify V and $\widehat{V}$. This we can do only because the isomorphism α_V is natural, in the sense that it is independent of the choice of basis.

Example 7.9

The mappings $f_1, f_2 : \mathbb{R}_1[X] \to \mathbb{R}$ given by

$$f_1(p) = \int_0^1 p, \quad f_2(p) = \int_0^2 p$$

are linear forms and, as is readily seen, $\{f_1, f_2\}$ forms a basis of the dual space $\mathbb{R}_1[X]^d$. Since $\mathbb{R}_1[X] \simeq \mathbb{R}^2$ under the correspondence

$$a_0 + a_1 X \longleftrightarrow (a_0, a_1)$$

we have that $\mathbb{R}_1[X]^d \simeq (\mathbb{R}^2)^d$. If $p = a_0 + a_1 X$ then

$$f_1(p) = a_0 + \tfrac{1}{2}a_1$$
$$f_2(p) = 2a_0 + 2a_1$$

and so we can associate with $\{f_1, f_2\}$ the basis $B = \{[1, \tfrac{1}{2}], [2, 2]\}$ of $(\mathbb{R}^2)^d$. By considering the matrix

$$P = \begin{bmatrix} 1 & 2 \\ \tfrac{1}{2} & 2 \end{bmatrix}$$

whose inverse is

$$P^{-1} = \begin{bmatrix} 2 & -2 \\ -\tfrac{1}{2} & 1 \end{bmatrix}$$

we see that $\{(2, -2), (-\tfrac{1}{2}, 1)\}$ is the basis of $\widehat{\mathbb{R}^2} = \mathbb{R}^2$ that is dual to B. Hence $\{2 - 2X, -\tfrac{1}{2} + X\}$ is the basis of $\mathbb{R}_1[X]$ that is dual to the basis $\{f_1, f_2\}$.

EXERCISES

7.6 Determine a basis of $\mathbb{R}^4$ whose dual basis is

$$\{[2, -1, 1, 0], [-1, 0, -2, 0], [-2, 2, 1, 0], [-8, 3, -3, 1]\}.$$

7.7 Let $\varphi_1, \varphi_2, \varphi_3$ be the linear forms on $\mathbb{R}_2[X]$ defined by

$$\varphi_1(p) = \int_0^1 p, \quad \varphi_2(p) = Dp(1), \quad \varphi_3(p) = p(0).$$

Show that $B = \{\varphi_1, \varphi_2, \varphi_3\}$ is a basis of $\mathbb{R}_2[X]^d$ and determine the basis $\{p_1, p_2, p_3\}$ of $\mathbb{R}_2[X]$ that is dual to B.

7.8 Let $\varphi_1, \ldots, \varphi_n \in (\mathbb{R}^n)^d$. Prove that the solution set C of the linear inequalities

$$\varphi_1(x) \geqslant 0, \ \varphi_2(x) \geqslant 0, \ \ldots, \ \varphi_n(x) \geqslant 0$$

satisfies the properties
(a) $\alpha, \beta \in C \Rightarrow \alpha + \beta \in C$;
(b) $\alpha \in C, t \in \mathbb{R}, t \geqslant 0 \Rightarrow t\alpha \in C$.

Show that if $\varphi_1, \ldots, \varphi_n$ form a basis of $(\mathbb{R}^n)^d$ then

$$C = \{t_1 \alpha_1 + \cdots + t_n \alpha_n \ ; \ t_i \in \mathbb{R}, t_i \geqslant 0\}$$

where $\{\alpha_1, \ldots, \alpha_n\}$ is the basis of $\mathbb{R}^n$ dual to the basis $\{\varphi_1, \ldots, \varphi_n\}$.

Suppose now that V, W are vector spaces over F and that $f : V \to W$ is linear. Given any $y^d \in W^d$ consider the linear mapping $y^d \circ f$. The diagram

$$V \xrightarrow{\ f\ } W \xrightarrow{\ y^d\ } F$$

shows that $y^d \circ f \in V^d$. Thus the assignment $y^d \mapsto y^d \circ f$ defines a mapping from W^d to V^d. We call this the **transpose** of f and denote it by f^t. Then we have

$$f^t(y^d) = y^d \circ f$$

and so f^t can be described as composition on the right by f. It is readily seen that $f^t \in \mathrm{Lin}(W^d, V^d)$.

- In terms of the notation introduced previously, we have

$$\langle f(x), y^d \rangle = y^d[f(x)] = (y^d \circ f)(x) = [f^t(y^d)](x) = \langle x, f^t(y^d) \rangle.$$

Consequently we have the identity

$$\boxed{\langle f(x), y^d \rangle = \langle x, f^t(y^d) \rangle}.$$

The above use of the word 'transpose' is suggested by the following result.

Theorem 7.4

Let V and W be vector spaces over F of dimensions m and n respectively. Let $(a_i)_m$, $(b_i)_n$ be ordered bases of V, W and let $f : V \to W$ be linear. If the matrix of f relative to $(a_i)_m$, $(b_i)_n$ is $A = [a_{ij}]_{n \times m}$ then relative to the corresponding dual ordered bases $(b_i^d)_n$, $(a_i^d)_m$ the matrix of f^t is A^t.

Proof

Let the matrix representing f^t be $B = [b_{ij}]_{m \times n}$. Consider the identity

$$\langle f(a_i), b_j^d \rangle = \langle a_i, f^t(b_j^d) \rangle.$$

The left hand side is

$$\left\langle \sum_{k=1}^{n} a_{ki} b_k,\ b_j^d \right\rangle = \sum_{k=1}^{n} a_{ki} \langle b_k, b_j^d \rangle = a_{ji}$$

whereas the right hand side is

$$\left\langle a_i,\ \sum_{k=1}^{n} b_{kj} a_k^d \right\rangle = \sum_{k=1}^{n} b_{kj} \langle a_i, a_k^d \rangle = b_{ij}.$$

Thus we have $b_{ij} = a_{ji}$ for all i, j and therefore $B = A^t$. $\square$

The main properties of transposition are the following.

Theorem 7.5

(1) $\mathrm{id}_V^t = \mathrm{id}_{V^d}$.

(2) *If* $f, g \in \mathrm{Lin}(V, W)$ *then* $(f + g)^t = f^t + g^t$.

(3) *If* $f \in \mathrm{Lin}(V, W)$ *and* $g \in \mathrm{Lin}(W, X)$ *then* $(g \circ f)^t = f^t \circ g^t$.

Proof

All are immediate from the fact that for every linear mapping h the effect of h' can be described as composition on the right by h. $\square$

Corollary

If $f : V \to W$ is an isomorphism then so is $f' : W^d \to V^d$, and $(f')^{-1} = (f^{-1})'$.

Proof

This follows from (1) and (3) on taking $g = f^{-1}$. $\square$

We can also consider the transpose of f'. We denote this by f'' and call it the **bitranspose** of f. There is a natural connection between bitransposes and biduals.

Theorem 7.6

For every linear mapping $f : V \to W$ the diagram

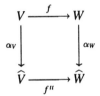

is commutative, in the sense that $f'' \circ \alpha_V = \alpha_W \circ f$.

Proof

We have to show that $f''(\hat{x}) = \widehat{f(x)}$ for every $x \in V$. Now for all $y^d \in V^d$ we have

$$[f''(\hat{x})](y^d) = (\hat{x} \circ f')(y^d) = \hat{x}[f'(y^d)]$$
$$= \langle x, f'(y^d) \rangle = \langle f(x), y^d \rangle = \widehat{f(x)}(y^d)$$

from which the result follows. $\square$

A consequence of Theorem 7.6 is that when V and W are of finite dimensions (in which case we agree to identify V, $\hat{V}$ and W, $\hat{W}$ and therefore also α_V, id_V and α_W, id_W) we have $f'' = f$. This then matches the matrix situation, where $A'' = A$.

EXERCISES

7.9 If V is a finite-dimensional vector space prove that a linear mapping $F : V \to V$ is surjective (resp. injective) if and only if its transpose is injective (resp. surjective).

7.10 Let V be a finite-dimensional vector space. Prove that the mapping
$$\varphi : \mathrm{Lin}(V, V) \to \mathrm{Lin}(V^d, V^d)$$
given by $\varphi(f) = f'$ is an isomorphism.

7.11 Let $f : \mathbb{R}^3 \to \mathbb{R}^3$ be given by the prescription

$$f(x, y, z) = (2x + y, \ x + y + z, \ -z).$$

Given the basis $X = \{(1,0,0), (1,1,0), (1,1,1)\}$ of $\mathbb{R}^3$ and the basis $Y^d = \{[1,0,0], [1,1,0], [1,1,1]\}$ of $(\mathbb{R}^3)^d$, determine the matrix of f' with respect to the bases Y^d and X^d.

7.12 Let $D : \mathbb{R}_n[X] \to \mathbb{R}$ be the differentiation mapping. Determine a basis for the kernel of the transpose of D.

Definition

If $x \in V$ and $y^d \in V^d$ are such that $\langle x, y^d \rangle = 0_V$ then we shall say that x is **annihilated** by y^d.

Since $\langle x, y^d \rangle = y^d(x)$ we see that the set of elements of V that are annihilated by y^d is Ker y^d. Now it is immediate from the identities (β) and (γ) preceding Theorem 7.3 that, for every non-empty subset E of V, the set of elements of V^d that annihilate every element of E is a subspace of V^d. We denote this subspace by E°. Thus

$$E^\circ = \{y^d \in V^d \ ; \ (\forall x \in E) \quad \langle x, y^d \rangle = 0\}.$$

We call E° the **annihilator** of E.

It is clear that $\{0_V\}^\circ = V^d$ and that $V^\circ = \{0_{V^d}\}$.

Theorem 7.7

If V is a finite-dimensional vector space and if W is a subspace of V then W° is also a subspace and

$$\dim W^\circ = \dim V - \dim W.$$

Moreover, identifying V and $\widehat{V}$, we have $W = W^{\circ\circ}$.

Proof

That W° is a subspace of V is clear. Suppose now that $\dim V = n$. The result is trivial if $W = V$ so suppose that $W \neq V$. Then $\dim W = m < n$. Let $\{a_1, \ldots, a_m\}$ be a basis of W and extend this to a basis

$$\{a_1, \ldots, a_m, a_{m+1}, \ldots, a_n\}$$

of V. Let $\{a_1^d, \ldots, a_n^d\}$ be the corresponding dual basis. If $x^d = \sum_{i=1}^{n} \lambda_i a_i^d \in W^\circ$ then for $j = 1, \ldots, m$ we have

$$0 = \langle a_j, x^d \rangle = \sum_{i=1}^{n} \lambda_i \langle a_j, a_i^d \rangle = \lambda_j.$$

It follows that $\{a_{m+1}^d, \ldots, a_n^d\}$ is a basis of W° and consequently

$$\dim W^\circ = n - m = \dim V - \dim W.$$

As for the final statement, consider the subspace $W^{\circ\circ} = (W^\circ)^\circ$ of $\widehat{V} = V$. By definition, every element of W is annihilated by every element of W° and so we have $W \subset W^{\circ\circ}$. On the other hand, by what we have just proved,

$$\dim W^{\circ\circ} = n - \dim W^\circ = n - (n - m) = m = \dim W.$$

It therefore follows that $W = W^{\circ\circ}$. $\square$

Annihilators and transposes are connected:

Theorem 7.8

If V and W are finite-dimensional and if $f : V \to W$ is linear then

(1) $(\operatorname{Im} f)^\circ = \operatorname{Ker} f^t$;
(2) $(\operatorname{Ker} f)^\circ = \operatorname{Im} f^t$;
(3) $\dim \operatorname{Im} f^t = \dim \operatorname{Im} f$;
(4) $\dim \operatorname{Ker} f^t = \dim \operatorname{Ker} f$.

Proof

(1) We have $y^d \in (\operatorname{Im} f)^\circ$ if and only if, for every $x \in V$,

$$0 = \langle f(x), y^d \rangle = \langle x, f^t(y^d) \rangle$$

which is the case if and only if $f^t(y^d) \in V^\circ = \{0_V\}$, i.e. if and only if $y^d \in \operatorname{Ker} f^t$.

(2) Replacing f by f^t in (1) and using the fact that $f^{tt} = f$, we obtain $(\operatorname{Im} f^t)^\circ = \operatorname{Ker} f$. Then, by Theorem 7.7, $(\operatorname{Ker} f)^\circ = (\operatorname{Im} f^t)^{\circ\circ} = \operatorname{Im} f^t$.

(3), (4) follow from (1), (2), and Theorem 6.7. $\square$

EXERCISES

7.13 Let V be a finite-dimensional vector space and let $(V_i)_{i \in I}$ be a family of subspaces of V. Prove that

$$\Big(\sum_{i \in I} V_i\Big)^\circ = \bigcap_{i \in I} V_i^\circ, \qquad \Big(\bigcap_{i \in I} V_i\Big)^\circ = \sum_{i \in I} V_i^\circ.$$

7.14 If V is a finite-dimensional vector space and W is a subspace of V, prove that $V^d/W^\circ \simeq W^d$.

[*Hint.* Show that $f - g \in W^\circ$ if and only the restrictions of f, g to W coincide.]

7.15 Prove that if S is a subset of a finite-dimensional vector space V then the subspace spanned by S is $S^{\circ\circ}$.

7.16 If W is the subspace of $\mathbb{R}^4$ spanned by $\{(1, 1, 0, 0), (0, 0, 1, 1)\}$, find a basis for W°.

7.17 Let A and B be subspaces of a vector space V such that $V = A \oplus B$ and let π_A, π_B be the projections onto A, B respectively. Prove that

$$V^d = \operatorname{Im} \pi_A^t \oplus \operatorname{Im} \pi_B^t.$$

7.18 By a **canonical isomorphism** $\zeta : V \to V^d$ we mean an isomorphism ζ such that, for all $x, y \in V$ and all isomorphisms $f : V \to V$, we have the identity

$$(\star) \qquad \langle x, \zeta(y) \rangle = \langle f(x), \zeta[f(y)] \rangle.$$

In this exercise we outline a proof of the fact that *if V is of dimension $n > 1$ over F then there is no canonical isomorphism $\zeta : V \to V^d$ except when $n = 2$ and $|F| = 2$.*

If ζ is such an isomorphism, show that if $y \neq 0_V$ then the subspace $\text{Ker } \zeta(y) = \{\zeta(y)\}^\circ$ is of dimension $n - 1$.

Suppose first that $n > 3$. If there exists $t \in \text{Ker } \zeta(t)$ for some $t \neq 0_V$, let $\{t, x_1, \dots, x_{n-2}\}$ be a basis of $\text{Ker } \zeta(t)$ and extend this to a basis $\{t, x_1, \dots, x_{n-2}, z\}$ of V. Let $f : V \to V$ be the linear mapping such that

$$f(t) = t, \quad f(x_1) = z, \quad f(z) = x_1, \quad f(x_i) = x_i \ (i \neq 1).$$

Show that f is an isomorphism that does not satisfy $(\star)$.
[*Hint.* Take $x = x_1, y = t$.]

If, on the other hand, $t \notin \text{Ker } \zeta(t)$ for all $t \neq 0_V$ let $\{x_1, \dots, x_{n-1}\}$ be a basis of $\text{Ker } \zeta(t)$ so that $\{x_1, \dots, x_{n-1}, t\}$ is a basis of V. Show that

$$\{x_1 + x_2, x_2, \dots, x_{n-1}, t\}$$

is also a basis of V. Show further that $x_2 \in \text{Ker } \zeta(x_1)$. Now show that if $f : V \to V$ is the linear mapping such that

$$f(x_1) = x_2, \quad f(x_2) = x_1 + x_2, \quad f(t) = t, \quad f(x_i) = x_i \ (i \neq 1, 2)$$

then f is an isomorphism that does not satisfy $(\star)$.

Conclude from these observations that we must have $n = 2$.

Suppose now that F has more than two elements and let $\lambda \in F$ be such that $\lambda \neq 0, 1$. If there exists $t \neq 0_V$ such that $t \in \text{Ker } \zeta(t)$, observe that $\{t\}$ is a basis of $\text{Ker } \zeta(t)$ and extend this to a basis $\{t, z\}$ of V. If $f : V \to V$ is the linear mapping such that

$$f(t) = t, \quad f(z) = \lambda z$$

show that f is an isomorphism that does not satisfy $(\star)$.
[*Hint.* Take $x = z, y = t$.]

If, on the other hand, $t \notin \text{Ker } \zeta(t)$ for all $t \neq 0_V$ let $\{z\}$ be a basis of $\text{Ker } \zeta(t)$ so that $\{t, z\}$ is a basis of V. If $f : V \to V$ is the linear mapping such that

$$f(z) = \lambda z, \quad f(t) = t$$

show that f is an isomorphism that does not satisfy $(\star)$.
[*Hint.* Take $x = y = z$.]

Conclude from these observations that $|F| = 2$.

Now examine the vector space F^2 where $F = \{0, 1\}$.

[*Hint.* $(F^2)^d$ is the set of linear mappings $f : F \times F \to F$. Since F^2 has four elements there are $2^4 = 16$ laws of composition on F. Only four of these are linear mappings from F^2 to F; and each of these is determined by its action on the natural basis of F^2. Compute $(F^2)^d$ and determine a canonical isomorphism from F^2 to $(F^2)^d$.]

We now pass to the consideration of the dual space of an inner product space. For this purpose, suppose that V is an inner product space and consider the inner product $(x, y) \mapsto \langle x \,|\, y \rangle$. For every $y \in V$ the assignment $x \mapsto \langle x \,|\, y \rangle$ is linear and therefore is an element of the dual space V^d. We shall write this element as y^d, so that we have the following useful amalgamated notation:

$$\boxed{\langle x \,|\, y \rangle = y^d(x) = \langle x, y^d \rangle}.$$

As we shall see, V and V^d are related via the following notion.

Definition

Let V and W be vector spaces over a field F where F is either $\mathbb{R}$ or $\mathbb{C}$. A mapping $f : V \to W$ is called a **conjugate transformation** if

$$(\forall x \in V)(\forall \lambda \in F) \qquad f(x + y) = f(x) + f(y), \quad f(\lambda x) = \overline{\lambda} f(x).$$

If, furthermore, f is a bijection then we say that it is a **conjugate isomorphism**.

Note that when $F = \mathbb{R}$ conjugate transformations are simply linear mappings.

Theorem 7.9

If V is a finite-dimensional inner product space then there is a conjugate isomorphism $\vartheta_V : V \to V^d$, namely that given by $\vartheta_V : x \mapsto x^d$.

Proof

For all $x, y, z \in V$ we have

$$\langle x, (y + z)^d \rangle = \langle x \,|\, y + z \rangle = \langle x \,|\, y \rangle + \langle x \,|\, z \rangle = \langle x, y^d \rangle + \langle x, z^d \rangle = \langle x, y^d + z^d \rangle$$

from which we obtain $(y + z)^d = y^d + z^d$, i.e. $\vartheta_V(y + z) = \vartheta_V(y) + \vartheta_V(z)$. Likewise,

$$\langle x, (\lambda y)^d \rangle = \langle x \,|\, \lambda y \rangle = \overline{\lambda} \langle x \,|\, y \rangle = \overline{\lambda} \langle x, y^d \rangle = \langle x, \overline{\lambda} y^d \rangle$$

and so $(\lambda y)^d = \overline{\lambda} y^d$, i.e. $\vartheta_V(\lambda y) = \overline{\lambda} \vartheta_V(y)$. Thus ϑ_V is a conjugate transformation.

That ϑ_V is injective follows from the fact that if $x \in \operatorname{Ker} \vartheta_V$ then $x^d = 0$ and so $\langle x \,|\, x \rangle = \langle x, x^d \rangle = 0$ whence $x = 0_V$.

To see that ϑ_V is also surjective, let $f \in V^d$. If $\{e_1, \ldots, e_n\}$ is an orthonormal basis of V, let

$$x = \sum_{i=1}^{n} \overline{f(e_i)} e_i.$$

Then for $j = 1, \ldots, n$ we have

$$x^d(e_j) = \langle e_j \mid x \rangle = \left\langle e_j \;\middle|\; \sum_{i=1}^{n} \overline{f(e_i)} e_i \right\rangle = \sum_{i=1}^{n} f(e_i)\langle e_j \mid e_i \rangle = f(e_j).$$

Thus x^d and f coincide on the basis $\{e_1, \ldots, e_n\}$. Hence $f = x^d = \vartheta_V(x)$ and so ϑ_V is surjective. Thus ϑ_V is a conjugate isomorphism. $\square$

Note from the above that we have the identity

$$\langle x \mid y \rangle = \langle x, \vartheta_V(y) \rangle.$$

Since ϑ_V is a bijection, we also have (writing $\vartheta_V^{-1}(y)$ for y) the identity

$$\langle x \mid \vartheta_V^{-1}(y) \rangle = \langle x, y \rangle.$$

We can now establish the following important result.

Theorem 7.10

Let V and W be finite-dimensional inner product spaces over the same field. Then for every linear mapping $f : V \to W$ there is a unique linear mapping $f^* : W \to V$ such that

$$(\forall x \in V)(\forall y \in W) \qquad \langle f(x) \mid y \rangle = \langle x \mid f^*(y) \rangle.$$

Proof

With the above notation, we have the equalities

$$\langle f(x) \mid y \rangle = \langle f(x), y^d \rangle = \langle x, f^t(y^d) \rangle$$
$$= \langle x \mid \vartheta_V^{-1}[f^t(y^d)] \rangle$$
$$= \langle x \mid (\vartheta_V^{-1} \circ f^t \circ \vartheta_W)(y) \rangle$$

from which it follows immediately that

$$f^* = \vartheta_V^{-1} \circ f^t \circ \vartheta_W$$

is the only linear mapping with the stated property. $\square$

Corollary

$f^* : W \to V$ is the unique linear mapping such that the diagram

$$
\begin{array}{ccc}
W & \xrightarrow{\ \vartheta_W\ } & W^d \\
{\scriptstyle f^*}\Big\downarrow & & \Big\downarrow{\scriptstyle f^t} \\
V & \xrightarrow[\ \vartheta_V\ ]{} & V^d
\end{array}
$$

is commutative, in the sense that $\vartheta_V \circ f^* = f^t \circ \vartheta_W$. $\square$

Definition

The unique linear mapping of Theorem 7.10 will be called the **adjoint** of f.

Example 7.10

Consider the vector space $V = \text{Mat}_{n \times n} \, \mathbb{C}$ with the inner product

$$\langle A \,|\, B \rangle = \text{tr } B^*A$$

where B^* denotes the transpose of the complex conjugate of B. For every $M \in V$ the mapping $\beta_M : V \to V$ given by $\beta_M(A) = MA$ is clearly linear. Since

$$\langle \beta_M(A) \,|\, B \rangle = \text{tr } B^*MA = \text{tr }(M^*B)^*A = \langle A \,|\, M^*B \rangle = \langle A \,|\, \beta_{M^*}(B) \rangle$$

we see that the adjoint of β_M exists and is β_{M^*}.

EXERCISES

7.19 Let V be a complex inner product space. For all $x, y \in V$ let $f_{x,y} : V \to V$ be given by

$$(\forall z \in V) \qquad f_{x,y}(z) = \langle z \,|\, y \rangle x.$$

Prove that f is linear and that

   (a) $(\forall x, y, z \in V) \; f_{x,y} \circ f_{y,z} = \|y\|^2 f_{x,z}$;

   (b)   the adjoint of $f_{x,y}$ is $f_{y,x}$.

7.20 If V is an inner product space then a linear mapping $f : V \to V$ is said to be **self-adjoint** if f^* exists and $f^* = f$; **normal** if f^* exists and $f \circ f^* = f^* \circ f$.

Prove that if $x \neq 0_V$ and $y \neq 0_V$ then the linear mapping $f_{x,y}$ of the previous exercise is

   (a)   normal if and only if there exists $\lambda \in \mathbb{C}$ such that $x = \lambda y$;

   (b)   self-adjoint if and only if there exists $\lambda \in \mathbb{R}$ such that $x = \lambda y$.

7.21 Let $C[0, 1]$ be the inner product space of real continuous functions on $[0, 1]$ with the integral inner product

$$\langle f \,|\, g \rangle = \int_0^1 fg.$$

Let $K : C[0, 1] \to C[0, 1]$ be the integral operator given by

$$K(f) = \int_0^1 xyf(y) \, dy.$$

Prove that K is self-adjoint.

7.22 If $f : \mathbb{R}^3 \to \mathbb{R}^3$ is given by

$$f(x, y, z) = (x + y, y, x + y + z),$$

determine f^*.

7.23 Let V be a finite-dimensional inner product space. Prove that for every $f \in V^d$ there is a unique $\beta \in V$ such that

$$(\forall x \in V) \qquad f(x) = \langle x \mid \beta \rangle.$$

[*Hint.* Let $\{\alpha_1, \ldots, \alpha_n\}$ be an orthonormal basis of V and consider the element $\beta = \sum_{i=1}^{n} \overline{f(\alpha_i)} \alpha_i.$]

Find such a β when $V = \mathbb{R}^3$ and $f(x, y, z) = x - 2y + 4z$.

7.24 Show as follows that the result of the previous exercise does not hold for an inner product space V of infinite dimension.

Let $z \in \mathbb{C}$ be fixed and define the 'evaluation at z' map $f_z \in V^d$ by

$$(\forall p \in V) \qquad f_z(p) = p(z).$$

Show that no $q \in V$ exists such that

$$(\forall p \in V) \; f_z(p) = \langle p \mid q \rangle.$$

[*Hint.* Suppose that such a q exists and let $r \in V$ be given by $r = X - z$. Show that, for every $p \in V$,

$$\int_0^1 r p \overline{q} = 0.$$

Now take $p = \overline{r} q$ and deduce the contradiction $q = 0$.]

7.25 Let $V = \mathbb{C}[X]$, the vector space over $\mathbb{C}$ of polynomials with complex coefficients. Show that

$$(p, q) \mapsto \langle p \mid q \rangle = \int_0^1 p \overline{q}$$

is an inner product on V.

If $p = \sum a_k X^k \in V$ define $\overline{p} = \sum \overline{a_k} X^k$. Let $f_p : V \to V$ be given by

$$(\forall q \in V) \qquad f_p(q) = pq.$$

Show that $f_p^\star$ exists and is $f_{\overline{p}}$.

Now let $D : V \to V$ be the differentiation mapping. Show that D does not admit an adjoint.

[*Hint.* Suppose that $D^\star$ exists and show that

$$\langle p \mid Dq + D^\star q \rangle = p(1)\overline{q}(1) - p(0)\overline{q}(0).$$

Suppose now that q is a fixed element of V such that $q(0) = 0$ and $q(1) = 1$. Use the previous exercise with $z = 1$ to obtain the required contradiction.]

The principal properties of the assignment $f \mapsto f^*$ are described in the following two results.

Theorem 7.11

Let V, W, X be finite-dimensional inner product spaces over the same field and let $f, g : V \to W$ and $h : W \to X$ be linear mappings. Then

(1) $(f + g)^* = f^* + g^*$;
(2) $(\lambda f)^* = \overline{\lambda} f^*$;
(3) $(h \circ f)^* = f^* \circ h^*$;
(4) $(f^*)^* = f$.

Proof

(1) is immediate from $f^* = \beta_V^{-1} \circ f^t \circ \beta_W$ and the fact that $(f + g)^t = f^t + g^t$.

(2) $\langle \lambda f(x) \mid y \rangle = \lambda \langle f(x) \mid y \rangle = \lambda \langle x \mid f^*(y) \rangle = \langle x \mid \overline{\lambda} f^*(y) \rangle$ and therefore, by the uniqueness of adjoints, $(\lambda f)^* = \overline{\lambda} f^*$.

(3) $\langle h[f(x)] \mid y \rangle = \langle f(x) \mid h^*(y) \rangle = \langle x \mid f^*[h^*(y)] \rangle$ and therefore, by the uniqueness of adjoints, $(h \circ f)^* = f^* \circ h^*$.

(4) Taking complex conjugates in the identity of Theorem 7.10 we obtain the identity

$$\langle f^*(y) \mid x \rangle = \langle y \mid f(x) \rangle,$$

from which it follows by the uniqueness of adjoints that $(f^*)^* = f$.   □

Theorem 7.12

Let V and W be finite-dimensional inner product spaces over the same field with $\dim V = \dim W$. If $f : V \to W$ is linear then the following are equivalent :

(1) *f is an inner product space isomorphism ;*
(2) *f is a vector space isomorphism and $f^{-1} = f^*$;*
(3) *$f \circ f^* = \mathrm{id}_W$;*
(4) *$f^* \circ f = \mathrm{id}_V$.*

Proof

(1) $\Rightarrow$ (2) : If (1) holds then f^{-1} exists and we have the identity

$$\langle f(x) \mid y \rangle = \langle f(x) \mid f[f^{-1}(y)] \rangle = \langle x \mid f^{-1}(y) \rangle$$

from which it follows by the uniqueness of adjoints that $f^{-1} = f^*$.

It is clear that (2) $\Rightarrow$ (3) and that (2) $\Rightarrow$ (4).

(4) $\Rightarrow$ (1) : If (4) holds then f is injective, hence bijective, and $f^{-1} = f^*$. Thus

$$(\forall x, y \in V) \qquad \langle f(x) \mid f(y) \rangle = \langle x \mid f^*[f(y)] \rangle = \langle x \mid y \rangle$$

and so f is an inner product isomorphism.

The proof of (3) $\Rightarrow$ (1) is similar.   □

EXERCISES

7.26 If V and W are finite-dimensional inner product spaces over the same field, prove that the assignment $f \mapsto f^*$ defines a conjugate isomorphism from $\mathrm{Lin}\,(V, W)$ to $\mathrm{Lin}\,(W, V)$.

7.27 Let V be a finite-dimensional inner product space and let $f : V \to V$ be linear. If W is a subspace of V, prove that W is f-invariant if and only if $W^\perp$ is f^*-invariant.

We have seen how the transpose f^t of a linear mapping f is such that $\mathrm{Ker}\,f^t$ and $\mathrm{Im}\,f^t$ are the annihilators of $\mathrm{Im}\,f$ and $\mathrm{Ker}\,f$ respectively. In view of the connection between transposes and adjoints, it will come as no surprise that $\mathrm{Ker}\,f^*$ and $\mathrm{Im}\,f^*$ are also related to the subspaces $\mathrm{Im}\,f$ and $\mathrm{Ker}\,f$.

Theorem 7.13

If V is a finite-dimensional inner product space and if $f : V \to V$ is linear then

$$\mathrm{Im}\,f^* = (\mathrm{Ker}\,f)^\perp, \quad \mathrm{Ker}\,f^* = (\mathrm{Im}\,f)^\perp.$$

Proof

Let $z \in \mathrm{Im}\,f^*$, say $z = f^*(y)$. Then for every $x \in \mathrm{Ker}\,f$ we have

$$\langle x \mid z \rangle = \langle x \mid f^*(y) \rangle = \langle f(x) \mid y \rangle = \langle 0_V \mid y \rangle = 0$$

and consequently $z \in (\mathrm{Ker}\,f)^\perp$. Thus $\mathrm{Im}\,f^* \subseteq (\mathrm{Ker}\,f)^\perp$.

Now let $y \in \mathrm{Ker}\,f^*$. Then for $z = f(x) \in \mathrm{Im}\,f$ we have

$$\langle z \mid y \rangle = \langle f(x) \mid y \rangle = \langle x, f^*(y) \rangle = \langle x \mid 0_V \rangle = 0$$

and consequently $y \in (\mathrm{Im}\,f)^\perp$. Thus $\mathrm{Ker}\,f^* \subseteq (\mathrm{Im}\,f)^\perp$.

Using Theorem 2.7, we then have

$$
\begin{aligned}
\dim \mathrm{Im}\,f &= \dim V - \dim(\mathrm{Im}\,f)^\perp \\
&\leqslant \dim V - \dim \mathrm{Ker}\,f^* \\
&= \dim \mathrm{Im}\,f^* \\
&\leqslant \dim(\mathrm{Ker}\,f)^\perp \\
&= \dim V - \dim \mathrm{Ker}\,f \\
&= \dim \mathrm{Im}\,f.
\end{aligned}
$$

The resulting equality gives

$$\dim \mathrm{Im}\,f^* = \dim(\mathrm{Ker}\,f)^\perp, \quad \dim(\mathrm{Im}\,f)^\perp = \dim \mathrm{Ker}\,f^*,$$

from which the result follows. $\quad\square$

We now investigate the natural question of how the matrices that represent f and f^* are related.

Definition

If $A = [a_{ij}]_{m \times n} \in \text{Mat}_{m \times n} \mathbb{C}$ then by the **adjoint** (or **conjugate transpose**) of A we mean the $n \times m$ matrix A^* such that $[A^*]_{ij} = \overline{a_{ji}}$.

Theorem 7.14

Let V and W be finite-dimensional inner product spaces over the same field. If, relative to ordered orthonormal bases $(v_i)_n$ and $(w_i)_m$ a linear mapping $f : V \to W$ is represented by the matrix A then the adjoint mapping $f^* : W \to V$ is represented, relative to the bases $(w_i)_m$ and $(v_i)_n$, by the adjoint matrix A^*.

Proof

For $j = 1, \ldots, n$ we have, by Theorem 1.5, $f(v_j) = \sum_{i=1}^{m} \langle f(v_j) | w_i \rangle w_i$. Thus, if $A = [a_{ij}]$ we have $a_{ij} = \langle f(v_j) | w_i \rangle$. Likewise, we have $f^*(w_j) = \sum_{i=1}^{n} \langle f^*(w_j) | v_i \rangle v_i$. Then, since

$$\overline{a_{ij}} = \overline{\langle f(v_j) | w_i \rangle} = \langle w_i | f(v_j) \rangle = \langle f^*(w_i) | v_j \rangle,$$

it follows that the matrix that represents f^* is A^*. $\square$

EXERCISE

7.28 If $A \in \text{Mat}_{n \times n} F$ prove that $\det A^* = \overline{\det A}$.

It is clear from Theorems 7.12 and 7.14 that a square matrix M represents an inner product space isomorphism if and only if M^{-1} exists and is M^*. Such a matrix is said to be **unitary**. It is readily seen by extending the corresponding results for ordinary vector spaces to inner product spaces that if A, B are $n \times n$ matrices over the ground field of V then A, B represent the same linear mapping with respect to possibly different ordered orthonormal bases of V if and only if there is a unitary matrix U such that $B = U^*AU = U^{-1}AU$. We describe this situation by saying that B is **unitarily similar** to A.

When the ground field is $\mathbb{R}$, the word **orthogonal** is often used instead of unitary. In this case M is orthogonal if and only if M^{-1} exists and is M^t, and B is said to be **orthogonally similar** to A.

It is clear that the relation of being unitarily (or orthogonally) similar is an equivalence relation on the set of $n \times n$ matrices over $\mathbb{C}$ (or $\mathbb{R}$). Just as with ordinary similarity, the problem of locating particularly simple representatives (or **canonical forms**) in certain equivalence classes is important from both the theoretical and practical points of view. We shall consider this problem later.

8
Orthogonal Direct Sums

In Theorem 2.6 we obtained, for an inner product space V and a finite-dimensional subspace W of V, a direct sum decomposition of the form $V = W \oplus W^{\perp}$. We now consider the following general notion.

Definition

Let $V_1, \ldots, V_n$ be non-zero subspaces of an inner product space V. Then V is said to be the **orthogonal direct sum** of $V_1, \ldots, V_n$ if

(1) $V = \bigoplus\limits_{i=1}^{n} V_i$;

(2) $(i = 1, \ldots, n) \quad V_i^{\perp} = \sum\limits_{j \neq i} V_j$.

In order to study orthogonal direct sum decompositions in an inner product space V let us begin by considering a projection $p : V \to V$ and the associated decomposition

$$V = \operatorname{Im} p \oplus \operatorname{Ker} p$$

as established in Theorem 2.4.

For this to be an orthogonal direct sum according to the above definition, it is clear that p has to be an **ortho-projection** in the sense that

$$\operatorname{Ker} p = (\operatorname{Im} p)^{\perp}, \quad \operatorname{Im} p = (\operatorname{Ker} p)^{\perp}.$$

Note that, by Theorem 2.7, when V is of finite dimension either of these conditions will do.

Example 8.1

As observed in Example 2.1, we have $\mathbb{R}^2 = Y \oplus D$ where $Y = \{(0, y) ; y \in \mathbb{R}\}$ and $D = \{(x, x) ; x \in \mathbb{R}\}$. Let p be the projection on D parallel to Y. Then for all $(x, y) \in \mathbb{R}^2$ we have $p(x, y) = (x, x)$, so that $\operatorname{Im} p = D$ and $\operatorname{Ker} p = Y$. But relative to the standard inner product on $\mathbb{R}^2$ we have $Y^{\perp} = \{(x, 0) ; x \in \mathbb{R}\}$. Hence $(\operatorname{Ker} p)^{\perp} \neq \operatorname{Im} p$ and so p is not an ortho-projection.

In order to discover precisely when a projection is an ortho-projection, we require the following result.

Theorem 8.1

If W and X are subspaces of a finite-dimensional inner product space V such that $V = W \oplus X$ then $V = W^{\perp} \oplus X^{\perp}$.

Proof

By Theorem 2.8, we have

$$\{0_V\} = V^{\perp} = (W + X)^{\perp} = W^{\perp} \cap X^{\perp}$$

and

$$V = \{0_V\}^{\perp} = (W \cap X)^{\perp} = W^{\perp} + X^{\perp}$$

whence we see that $V = W^{\perp} \oplus X^{\perp}$. $\square$

Corollary

If $V = W \oplus X$ and if p is the projection on W parallel to X then $p^{\star}$ is the projection on $X^{\perp}$ parallel to $W^{\perp}$.

Proof

Since p is idempotent it follows by Theorem 7.11(3) that $p^{\star} \circ p^{\star} = (p \circ p)^{\star} = p^{\star}$. Hence $p^{\star}$ is idempotent and so is the projection on $\operatorname{Im} p^{\star}$ parallel to $\operatorname{Ker} p^{\star}$. By Theorem 2.3, $\operatorname{Im} p = W$ and $\operatorname{Ker} p = X$, so by Theorem 7.13 we have $W^{\perp} = (\operatorname{Im} p)^{\perp} = \operatorname{Ker} p^{\star}$ and $X^{\perp} = (\operatorname{Ker} p)^{\perp} = \operatorname{Im} p^{\star}$. $\square$

Definition

If V is an inner product space then a linear mapping $f : V \to V$ is said to be **self-adjoint** if $f = f^{\star}$.

We can now characterise ortho-projections.

Theorem 8.2

Let V be a finite-dimensional inner product space. If $p : V \to V$ is a projection then p is an ortho-projection if and only if p is self-adjoint.

Proof

By the Corollary of Theorem 8.1, $p^{\star}$ is the projection on $\operatorname{Im} p^{\star} = (\operatorname{Ker} p)^{\perp}$ parallel to $\operatorname{Ker} p^{\star} = (\operatorname{Im} p)^{\perp}$. Thus, if p is an ortho-projection we must have $\operatorname{Im} p^{\star} = \operatorname{Im} p$. It then follows by Theorem 2.3 that for every $x \in V$ we have $p(x) = p^{\star}[p(x)]$. Consequently $p = p^{\star} \circ p$ and therefore, by Theorem 7.11(3)(4),

$$p^{\star} = (p^{\star} \circ p)^{\star} = p^{\star} \circ p^{\star\star} = p^{\star} \circ p = p.$$

Hence p is self-adjoint.

Conversely, if $p = p^*$ then clearly $\operatorname{Im} p = \operatorname{Im} p^* = (\operatorname{Ker} p)^\perp$, which shows that p is an ortho-projection. □

If now W is a finite-dimensional subspace of an inner product space V then by Theorem 2.6 we have $V = W \oplus W^\perp$. Now Theorem 8.1 and its corollary guarantee the existence of an ortho-projection on W. There is in fact precisely one ortho-projection on W. To see this, suppose that p and q are ortho-projections on W. Then $\operatorname{Im} p = W = \operatorname{Im} q$ and $\operatorname{Ker} p = (\operatorname{Im} p)^\perp = (\operatorname{Im} q)^\perp = \operatorname{Ker} q$. Since projections are uniquely determined by their images and kernels, it follows that $p = q$. We may therefore talk of *the* ortho-projection on W. It may be characterised as follows.

Theorem 8.3

Let W be a finite-dimensional subspace of an inner product space V. If p_W is the ortho-projection on W then for every $x \in V$ the element $p_W(x)$ is the unique element of W that is nearest x, in the sense that

$$(\forall w \in W) \quad \|x - p_W(x)\| \leqslant \|x - w\|.$$

Proof

Since p_W is idempotent, for every $w \in W$ we have that $x - p_W(x) \in \operatorname{Ker} p_W = W^\perp$. Consequently,

$$\|x - w\|^2 = \|p_W(x) - w + x - p_W(x)\|^2$$
$$= \|p_W(x) - w\|^2 + \|x - p_W(x)\|^2$$
$$\geqslant \|x - p_W(x)\|^2$$

whence the required inequality follows. □

Corollary

If $\{e_1, \dots, e_n\}$ is an orthonormal basis of W then

$$(\forall x \in V) \quad p_W(x) = \sum_{i=1}^{n} \langle x \mid e_i \rangle e_i.$$

Proof

This is immediate from the above and the discussion that precedes Example 2.9. □

EXERCISES

8.1 In $\mathbb{R}^4$ consider the subspace $W = \operatorname{Span}\{(0, 1, 2, 0), (1, 0, 0, 1)\}$. Determine the ortho-projection of $(1, 2, 0, 0)$ on W.

8.2 Let W be a finite-dimensional subspace of an inner product space V. Prove that $p_{W^\perp} = \operatorname{id}_V - p_W$.

8.3 Let V be a finite-dimensional inner product space. If W is a subspace of V, prove that

$$(\forall x \in V) \quad \|p_W(x)\| \leqslant \|x\|.$$

8.4 In $\mathbb{R}^2$ consider a line L passing through the origin and making an angle ϑ with the x-axis. For every point (x, y) we have that $p_L(x, y) = (x^\star, y^\star)$ where $(x^\star, y^\star)$ is the foot of the perpendicular from (x, y) to L. Show by simple geometrical considerations that

$$\begin{bmatrix} x^\star \\ y^\star \end{bmatrix} = \begin{bmatrix} \cos^2 \vartheta & \sin \vartheta \cos \vartheta \\ \sin \vartheta \cos \vartheta & \sin^2 \vartheta \end{bmatrix} \begin{bmatrix} x \\ y \end{bmatrix}.$$

8.5 Consider the linear mapping $f : \mathbb{R}^3 \to \mathbb{R}^3$ given by

$$f(x, y, z) = (-z, x + z, y + z).$$

Determine the ortho-projections on the eigenspaces of f.

Concerning direct sums of subspaces in an inner product space we have the following result.

Theorem 8.4

Let V be a finite-dimensional inner product space and let $V_1, \ldots, V_k$ be subspaces of V such that $V = \bigoplus_{i=1}^{k} V_i$. Then this direct sum is an orthogonal direct sum if and only if, for each i, every element of V_i is orthogonal to every element of V_j $(j \neq i)$.

Proof

The necessity is clear. As for sufficiency, if every element of V_i is orthogonal to every element of V_j $(j \neq i)$ then we have $\sum_{j \neq i} V_j \subseteq V_i^\perp$. But

$$\dim \sum_{j \neq i} V_j = \dim V - \dim V_i = \dim V_i^\perp.$$

Consequently $\sum_{j \neq i} V_j = V_i^\perp$. $\square$

Definition

Let V be a finite-dimensional inner product space and let $f : V \to V$ be linear. Then we say that f is **ortho-diagonalisable** if there is an orthonormal basis of V that consists of eigenvectors of f; equivalently, if there is an orthonormal basis of V with respect to which the matrix of f is diagonal.

Our objective now is to determine under what conditions a linear mapping is ortho-diagonalisable. In purely matrix terms, this problem is that of determining when a given square matrix (over $\mathbb{R}$ or $\mathbb{C}$) is unitarily similar to a diagonal matrix.

For this purpose, we recall that if f is diagonalisable then by Theorem 3.3 its minimum polynomial is

$$m_f = (X - \lambda_1)(X - \lambda_2) \cdots (X - \lambda_k)$$

where $\lambda_1, \ldots, \lambda_k$ are the distinct eigenvalues of f. Moreover, by Corollary 2 of Theorem 3.2 we have that $V = \bigoplus_{i=1}^{k} V_{\lambda_i}$ where the subspace $V_{\lambda_i} = \mathrm{Ker}(f - \lambda_i \mathrm{id}_V)$ is the eigenspace associated with λ_i. If now $p_i : V \to V$ is the projection on V_{λ_i} parallel to $\sum_{j \neq i} V_{\lambda_j}$ then from Theorem 2.5 we have $\sum_{i=1}^{k} p_i = \mathrm{id}_V$ and $p_i \circ p_j = 0$ when $i \neq j$. Consequently, for every $x \in V$, we see that

$$f(x) = f\left(\sum_{i=1}^{k} p_i(x)\right) = \sum_{i=1}^{k} f[p_i(x)] = \sum_{i=1}^{k} \lambda_i p_i(x) = \left(\sum_{i=1}^{k} \lambda_i p_i\right)(x),$$

which gives $f = \sum_{i=1}^{k} \lambda_i p_i$.

As we shall now show, if V is an inner product space then the condition that each p_i be self-adjoint is the key to f being ortho-diagonalisable.

Theorem 8.5

Let V be a non-zero finite-dimensional inner product space and let $f : V \to V$ be linear. Then f is ortho-diagonalisable if and only if there are non-zero self-adjoint projections $p_1, \ldots, p_k : V \to V$ and distinct scalars $\lambda_1, \ldots, \lambda_k$ such that

(1) $f = \sum_{i=1}^{k} \lambda_i p_i$;

(2) $\sum_{i=1}^{k} p_i = \mathrm{id}_V$;

(3) $(i \neq j) \quad p_i \circ p_j = 0.$

Proof

$\Rightarrow$: It suffices to note that if $\bigoplus_{i=1}^{k} V_{\lambda_i}$ is an orthogonal direct sum then each p_i is an ortho-projection and so, by Theorem 8.3, is self-adjoint.

$\Leftarrow$: If the conditions hold then by Theorem 2.5 we have $V = \bigoplus_{i=1}^{k} \mathrm{Im}\, p_i$. Now the λ_i that appear in (1) are the distinct eigenvalues of f. To see this, observe that

$$f \circ p_j = \left(\sum_{i=1}^{k} \lambda_i p_i\right) \circ p_j = \sum_{i=1}^{k} \lambda_i (p_i \circ p_j) = \lambda_j p_j$$

and so $(f - \lambda_j \mathrm{id}_V) \circ p_j = 0$ whence

$$\{0_V\} \neq \mathrm{Im}\, p_j \subseteq \mathrm{Ker}(f - \lambda_j \mathrm{id}_V).$$

Thus each λ_j is an eigenvalue of f.

On the other hand, for every scalar λ,

$$f - \lambda \mathrm{id}_V = \sum_{i=1}^{k} \lambda_i p_i - \sum_{i=1}^{k} \lambda p_i = \sum_{i=1}^{k} (\lambda_i - \lambda) p_i,$$

so that, if x is an eigenvector of f associated with the eigenvalue λ, we have

$$\sum_{i=1}^{k} (\lambda_i - \lambda) p_i(x) = 0_V.$$

But $V = \bigoplus_{i=1}^{k} \mathrm{Im}\, p_i$, and so we deduce that $(\lambda_i - \lambda) p_i(x) = 0_V$ for $i = 1, \ldots, k$. If now $\lambda \neq \lambda_i$ for every i then we would have $p_i(x) = 0_V$ for every i, whence there would follow the contradiction $x = \sum_{i=1}^{k} p_i(x) = 0_V$. Consequently $\lambda = \lambda_i$ for some i. Hence $\lambda_1, \ldots, \lambda_k$ are the distinct eigenvalues of f.

We now observe that, for each j,

$$\mathrm{Im}\, p_j = \mathrm{Ker}\,(f - \lambda_j \mathrm{id}_V).$$

For, suppose that $x \in \mathrm{Ker}\,(f - \lambda_j \mathrm{id}_V)$, so that $f(x) = \lambda_j x$. Then $0_V = \sum_{i=1}^{k} (\lambda_i - \lambda_j) p_i(x)$ and therefore $(\lambda_i - \lambda_j) p_i(x) = 0_V$ for all i whence $p_i(x) = 0_V$ for all $i \neq j$. Then

$$x = \sum_{i=1}^{k} p_i(x) = p_j(x) \in \mathrm{Im}\, p_j$$

and so $\mathrm{Ker}\,(f - \lambda_j \mathrm{id}_V) \subseteq \mathrm{Im}\, p_j$. The reverse inclusion was established above.

Since $V = \bigoplus_{i=1}^{k} \mathrm{Im}\, p_i = \bigoplus_{i=1}^{k} \mathrm{Ker}\,(f - \lambda_i \mathrm{id}_V)$ it follows that V has a basis consisting of eigenvectors of f and so f is diagonalisable.

Now by hypothesis the projections p_i are self-adjoint so, for $j \neq i$,

$$\langle p_i(x) \,|\, p_j(x) \rangle = \langle p_i(x) \,|\, p_j^\star(x) \rangle = \langle p_j[p_i(x)] \,|\, x \rangle = \langle 0_V \,|\, x \rangle = 0.$$

It follows that the above basis of eigenvectors is orthogonal. By normalising each vector in this basis we obtain an orthonormal basis of eigenvectors. Hence f is ortho-diagonalisable. $\square$

Definition

For an ortho-diagonalisable mapping f the equality $f = \sum_{i=1}^{k} \lambda_i p_i$ of Theorem 8.5 is called the **spectral resolution** of f.

Applying the results of Theorem 7.11 to the conditions in Theorem 8.5 we obtain

$$(1^\star) \quad f^\star = \sum_{i=1}^{k} \overline{\lambda_i} p_i^\star = \sum_{i=1}^{k} \overline{\lambda_i} p_i; \qquad (2^\star) = (2), \qquad (3^\star) = (3).$$

It follows immediately by Theorem 8.5 that f^* is also ortho-diagonalisable and that (1*) gives its spectral resolution. Note as a consequence that $\overline{\lambda_1}, \ldots, \overline{\lambda_k}$ are the distinct eigenvalues of f^*.

A simple calculation now reveals that

$$f \circ f^* = \sum_{i=1}^{k} |\lambda_i|^2 p_i = f^* \circ f$$

from which we deduce that ortho-diagonalisable mappings commute with their adjoints.

This observation leads to the following notion.

Definition

If V is a finite-dimensional inner product space and if $f : V \to V$ is linear then we say that f is **normal** if it commutes with its adjoint. Similarly, a square matrix A over a field is said to be **normal** if $AA^* = A^*A$.

Example 8.2

For the matrix $A \in \text{Mat}_{2\times2} \, \mathbb{C}$ given by

$$A = \begin{bmatrix} 2 & i \\ 1 & 2 \end{bmatrix}$$

we have

$$A^* = \begin{bmatrix} 2 & 1 \\ -i & 2 \end{bmatrix}.$$

Then

$$AA^* = \begin{bmatrix} 5 & 2+2i \\ 2-2i & 5 \end{bmatrix} = A^*A$$

and so A is normal.

EXERCISES

8.6 If the matrix A is normal and non-singular prove that so also is A^{-1}.

8.7 If A and B are real symmetric matrices prove that $A + iB$ is normal if and only if A and B commute.

8.8 Show that each of the matrices

$$A = \begin{bmatrix} 1 & i \\ -i & 1 \end{bmatrix}, \quad B = \begin{bmatrix} 0 & i \\ i & 0 \end{bmatrix}$$

is normal, but neither $A + B$ nor AB is normal.

8.9 Let V be a complex inner product space and let $f : V \to V$ be linear. Define

$$f_1 = \tfrac{1}{2}(f + f^\star), \quad f_2 = \tfrac{1}{2i}(f - f^\star).$$

Prove that

(1) f_1 and f_2 are self-adjoint;

(2) $f = f_1 + if_2$;

(3) if $f = g_1 + ig_2$ where g_1 and g_2 are self-adjoint then $g_1 = f_1$ and $g_2 = f_2$;

(4) f is normal if and only if f_1 and f_2 commute.

8.10 Prove that the linear mapping $f : \mathbb{C}^2 \to \mathbb{C}^2$ given by

$$f(\alpha, \beta) = (2\alpha + i\beta, \alpha + 2\beta)$$

is normal and not self-adjoint.

8.11 Let V be a finite-dimensional inner product space and let $f : V \to V$ be linear. Prove that if $\alpha, \beta \in \mathbb{C}$ are such that $|\alpha| = |\beta|$ then $\alpha f + \beta f^\star$ is normal.

8.12 Show that $A \in \mathrm{Mat}_{3\times 3}\,\mathbb{C}$ given by

$$A = \begin{bmatrix} 1+i & -1 & 2-i \\ -i & 2+2i & 0 \\ -1+2i & 0 & 3+3i \end{bmatrix}$$

is normal.

We have seen above that a necessary condition for a linear mapping to be ortho-diagonalisable is that it be normal. It is quite remarkable that when the ground field is $\mathbb{C}$ this condition is also sufficient. In order to establish this, we require the following properties of normal mappings.

Theorem 8.6

Let V be a non-zero finite-dimensional inner product space and let $f : V \to V$ be a linear mapping that is normal. Then

(1) $(\forall x \in V) \quad \|f(x)\| = \|f^\star(x)\|$;

(2) $p(f)$ *is normal for every polynomial p;*

(3) $\mathrm{Im}\,f \cap \mathrm{Ker}\,f = \{0_V\}$.

Proof

(1) Since $f \circ f^\star = f^\star \circ f$ we have, for all $x \in V$,

$$\langle f(x)\,|\,f(x)\rangle = \langle x\,|\,f^\star[f(x)]\rangle = \langle x\,|\,f[f^\star(x)]\rangle = \langle f^\star(x)\,|\,f^\star(x)\rangle$$

from which (1) follows.

(2) If $p = \sum_{i=0}^{n} a_i X^i$ then we have $p(f) = \sum_{i=0}^{n} a_i f^i$ and it follows by Theorem 7.11

that $[p(f)]^* = \sum_{i=0}^{n} \bar{a}_i (f^*)^i$. Since f and f^* commute, it follows that so do $p(f)$ and
$[p(f)]^*$. Hence $p(f)$ is normal.

(3) If $x \in \text{Im} f \cap \text{Ker} f$ then there exists $y \in V$ such that $x = f(y)$ and $f(x) = 0_V$.
By (1) we have $f^*(x) = 0_V$ and therefore

$$0 = \langle f^*(x) \mid y \rangle = \langle x \mid f(y) \rangle = \langle x \mid x \rangle$$

whence $x = 0_V$. $\quad\square$

We now characterise the normal projections.

Theorem 8.7

*Let V be a non-zero finite-dimensional inner product space. If $p : V \rightarrow V$ is a
projection then p is normal if and only if it is self-adjoint.*

Proof

Clearly, if p is self-adjoint then p is normal. Conversely, suppose that p is normal. By
Theorem 8.6 we have $\|p(x)\| = \|p^*(x)\|$ for every $x \in V$ and so $p(x) = 0_V$ if and only
if $p^*(x) = 0_V$. Now, given $x \in V$, let $y = x - p(x)$. We have $p(y) = p(x) - p(x) = 0_V$
and so $0_V = p^*(y) = p^*(x) - p^*[p(x)]$ which gives $p^* = p^* \circ p$. Thus, by Theorem
7.11, $p = p^{**} = (p^* \circ p)^* = p^* \circ p^{**} = p^* \circ p = p^*$, i.e. p is self-adjoint. $\quad\square$

We can now solve the ortho-diagonalisation problem for **complex** inner product
spaces.

Theorem 8.8

*Let V be a non-zero finite-dimensional complex inner product space. If $f : V \rightarrow V$
is linear then f is ortho-diagonalisable if and only if it is normal.*

Proof

We have already seen that the condition is necessary. As for sufficiency, suppose
that f is normal. To show first that f is diagonalisable it suffices to show that its
minimum polynomial m_f is a product of distinct linear factors. For this purpose, we
shall make use of the fact that the field $\mathbb{C}$ of complex numbers is algebraically closed,
in the sense that every polynomial of degree at least 1 can be expressed as a product
of linear polynomials. Thus m_f is certainly a product of linear factors. Suppose, by
way of obtaining a contradiction, that $\alpha \in \mathbb{C}$ is such that $X - \alpha$ is a multiple factor
of m_f. Then $m_f = (X - \alpha)^2 g$ for some polynomial g. Thus for every $x \in V$ we have

$$0_V = [m_f(f)](x) = [(f - \alpha \text{id}_V)^2 \circ g(f)](x)$$

and consequently $[(f - \alpha \text{id}_V) \circ g(f)](x)$ belongs to both the image and the kernel
of $f - \alpha \text{id}_V$. Since, by Theorem 8.6(2), $f - \text{id}_V$ is normal we deduce from Theorem
8.6(3) that

$$(\forall x \in V) \qquad [(f - \alpha \text{id}_V) \circ g(f)](x) = 0_V.$$

Consequently $(f - \alpha \mathrm{id}_V) \circ g(f)$ is the zero mapping on V, so f is a zero of the polynomial $(X - \alpha)g$. This contradicts the fact that $(X - \alpha)^2 g$ is the minimum polynomial of f. Thus we see that f is diagonalisable.

To show that f is ortho-diagonalisable, it suffices to show that the corresponding projections p_i onto the eigenspaces are ortho-projections, and by Theorem 8.2 it is enough to show that they are self-adjoint.

Now since f is diagonalisable we have $f = \sum_{i=1}^{k} \lambda_i p_i$. Since $p_i \circ p_j = 0$ for $i \neq j$, it is

readily seen that $f^2 = \sum_{i=1}^{k} \lambda_i^2 p_i$. A simple inductive argument now gives $f^n = \sum_{i=1}^{k} \lambda_i^n p_i$ for all $n \geqslant 1$. An immediate consequence of this is that for every polynomial g we have

$$g(f) = \sum_{i=1}^{k} g(\lambda_i) p_i.$$

Consider now the Lagrange polynomials $L_1, \ldots, L_k$ associated with $\lambda_1, \ldots, \lambda_k$. We recall (see Example 7.6) that

$$L_j = \prod_{i \neq j} \frac{X - \lambda_i}{\lambda_j - \lambda_i}.$$

Taking $g = L_j$ we see from the above that

$$L_j(f) = \sum_{i=1}^{k} L_j(\lambda_i) p_i = \sum_{i=1}^{k} \delta_{ij} p_i = p_j.$$

It therefore follows by Theorem 8.6(2) that each p_j is normal and so, by Theorem 8.7, is self-adjoint. $\square$

Corollary

A square matrix over $\mathbb{C}$ is unitarily similar to a diagonal matrix if and only if it is normal. $\square$

EXERCISES

8.13 If A is a normal matrix show that so also is $g(A)$ for every polynomial g.

8.14 Prove that $A^* = p(A)$ for some polynomial p if and only if A is normal.

8.15 For each of the following linear mappings determine whether or not it is ortho-diagonalisable:

(1) $f : \mathbb{C}^2 \to \mathbb{C}^2$; $f(x, y) = (x + iy, -ix + y)$.

(2) $f : \mathbb{C}^2 \to \mathbb{C}^2$; $f(x, y) = (x + (1 + i)y, (1 - i)x + y)$.

(3) $f : \mathbb{C}^3 \to \mathbb{C}^3$; $f(x, y, z) = (y, z, i(x + y + z))$.

(4) $f : \mathbb{C}^3 \to \mathbb{C}^3$; $f(x, y, z) = (x + iy, -ix + y + (1 - i)z, (1 + i)y + z)$.

It should be noted that in the proof of Theorem 8.8 we made use of the fact that the field $\mathbb{C}$ of complex numbers is algebraically closed. This is not true of $\mathbb{R}$ and so we might expect that the corresponding result fails in general for real inner product spaces (and real square matrices). This is indeed the case: there exist normal linear mappings on a real inner product space that are not diagonalisable. One way in which this can happen is when all the eigenvalues of the mapping in question are complex.

Example 8.3

In the real inner product space $\mathbb{R}^2$ consider the matrix

$$R_{2\pi/3} = \begin{bmatrix} -\frac{1}{2} & -\frac{\sqrt{3}}{2} \\ \frac{\sqrt{3}}{2} & -\frac{1}{2} \end{bmatrix}$$

which represents an anti-clockwise rotation through an angle $2\pi/3$ of the coordinate axes. It is readily seen that $R_{2\pi/3}$ is normal. But its minimum polynomial is $X^2 + X + 1$, and this has no zeros in $\mathbb{R}$. Hence this matrix is not diagonalisable over $\mathbb{R}$.

Thus, in order to obtain an analogue of Theorem 8.8 in the case where the ground field is $\mathbb{R}$ we are led first to the consideration of normal linear mappings whose eigenvalues are all real. These can be characterised as follows.

Theorem 8.9

Let V be a non-zero finite-dimensional complex inner product space. If $f : V \to V$ is linear then the following conditions are equivalent:

(1) *f is normal and all its eigenvalues are real;*
(2) *f is self-adjoint.*

Proof

(1) $\Rightarrow$ (2) : By Theorem 8.8, f is ortho-diagonalisable. Let $f = \sum_{i=1}^{k} \lambda_i p_i$ be its spectral resolution. We know that f^* is also normal, with spectral resolution $f^* = \sum_{i=1}^{k} \overline{\lambda_i} p_i$. Since each λ_i is real, it follows that $f^* = f$ and so f is self-adjoint.

(2) $\Rightarrow$ (1) : If (2) holds then f is normal. If then $f = \sum_{i=1}^{k} \lambda_i p_i$ and $f^* = \sum_{i=1}^{k} \overline{\lambda_i} p_i$ are the spectral resolutions of f and f^* then $f = f^*$ gives $\sum_{i=1}^{k} (\lambda_i - \overline{\lambda_i}) p_i = 0$ and so

$\sum_{i=1}^{k} (\lambda_i - \overline{\lambda_i}) p_i(x) = 0_V$ for every $x \in V$, whence $(\lambda_i - \overline{\lambda_i}) p_i = 0$ for every i since $V = \bigoplus_{i=1}^{k} \operatorname{Im} p_i$. Since none of the p_i can be zero, we deduce that $\lambda_i = \overline{\lambda_i}$ for every i. Thus every eigenvalue of f is real. $\square$

Corollary

All the eigenvalues of a self-adjoint matrix are real. □

EXERCISES

8.16 A complex matrix A is such that $A^*A = -A$. Prove that A is self-adjoint and that its eigenvalues are either 0 or −1.

8.17 Prove that a complex matrix is unitary $(A^* = A^{-1})$ if and only if it is normal and every eigenvalue is of modulus 1.

The analogue of Theorem 8.8 for **real** inner product spaces is the following.

Theorem 8.10

Let V be a non-zero finite-dimensional real inner product space. If $f : V \to V$ is linear then f is ortho-diagonalisable if and only if it is self-adjoint.

Proof

$\Rightarrow$: If f is ortho-diagonalisable let $f = \sum_{i=1}^{k} \lambda_i p_i$ be its spectral resolution. Since the ground field is $\mathbb{R}$, every λ_i is real. Thus, taking adjoints and using Theorem 8.2, we obtain $f^* = f$.

$\Leftarrow$: Conversely, suppose that f is self-adjoint and let A be the $(n \times n$ say) matrix of f relative to some ordered orthonormal basis of V. Then clearly A is symmetric. Now let f^+ be the linear mapping on the complex inner product space $\mathbb{C}^n$ whose matrix relative to the standard (orthonormal) basis is A. Then f^+ is self-adjoint. By Theorem 8.9, the eigenvalues of f^+ are all real and, since f^+ is diagonalisable, the minimum polynomial of f^+ is a product of distinct linear polynomials over $\mathbb{R}$. Since this is then the minimum polynomial of A, it is also the minimum polynomial of f. Thus we conclude that f is diagonalisable. That it is in fact ortho-diagonalisable is established precisely as in the proof of Theorem 8.7. □

Corollary

If A is a square matrix over $\mathbb{R}$ then A is orthogonally similar to a diagonal matrix if and only if A is symmetric. □

Example 8.4

By the above corollary, the real symmetric matrix

$$A = \begin{bmatrix} 4 & 2 & 2 \\ 2 & 4 & 2 \\ 2 & 2 & 4 \end{bmatrix}$$

is orthogonally similar to a diagonal matrix D. Let us find an orthogonal matrix P such that $P^{-1}AP = D$.

It is readily seen that the characteristic polynomial of A is

$$c_A = (X - 2)^2 (X - 8),$$

so the eigenvalues of A are 2 and 8. A basis for the eigenspace associated with the eigenvalue 2 is

$$\left\{ \begin{bmatrix} -1 \\ 1 \\ 0 \end{bmatrix}, \begin{bmatrix} -1 \\ 0 \\ 1 \end{bmatrix} \right\}.$$

Applying the Gram–Schmidt process, we see that an orthonormal basis for this eigenspace is

$$\left\{ \begin{bmatrix} -\frac{1}{\sqrt{2}} \\ \frac{1}{\sqrt{2}} \\ 0 \end{bmatrix}, \begin{bmatrix} -\frac{1}{\sqrt{6}} \\ -\frac{1}{\sqrt{6}} \\ \frac{2}{\sqrt{6}} \end{bmatrix} \right\}.$$

As for the eigenspace associated with the eigenvalue 8, a basis is

$$\left\{ \begin{bmatrix} 1 \\ 1 \\ 1 \end{bmatrix} \right\},$$

so an orthonormal basis of this eigenspace is

$$\left\{ \begin{bmatrix} \frac{1}{\sqrt{3}} \\ \frac{1}{\sqrt{3}} \\ \frac{1}{\sqrt{3}} \end{bmatrix} \right\}.$$

Pasting these orthonormal bases together, we see that the matrix

$$P = \begin{bmatrix} -\frac{1}{\sqrt{2}} & -\frac{1}{\sqrt{6}} & \frac{1}{\sqrt{3}} \\ \frac{1}{\sqrt{2}} & -\frac{1}{\sqrt{6}} & \frac{1}{\sqrt{3}} \\ 0 & \frac{2}{\sqrt{6}} & \frac{1}{\sqrt{3}} \end{bmatrix}$$

is orthogonal and such that $P^{-1}AP = \text{diag}\{2, 2, 8\}$.

EXERCISE

8.18 Find an orthogonal matrix P such that $P^{-1}AP$ is diagonal where

$$A = \begin{bmatrix} 4\frac{1}{2} & 0 & -3\frac{1}{2} \\ 0 & -1 & 0 \\ -3\frac{1}{2} & 0 & 4\frac{1}{2} \end{bmatrix}.$$

Hence find a matrix B such that $B^3 = A$.

We shall now derive a useful alternative characterisation of a self-adjoint linear mapping on a complex inner product space. In order to do so, we require the following result.

Theorem 8.11

Let V be a complex inner product space. If $f : V \to V$ is linear and such that $\langle f(x) \,|\, x \rangle = 0$ for all $x \in V$ then $f = 0$.

Proof

Using the condition we see that, for all $y, z \in V$,

$$0 = \langle f(y+z) \,|\, y+z \rangle = \langle f(y) \,|\, z \rangle + \langle f(z) \,|\, y \rangle;$$
$$0 = \langle f(iy+z) \,|\, iy+z \rangle = i\langle f(y) \,|\, z \rangle - i\langle f(z) \,|\, y \rangle,$$

from which it follows that $\langle f(y) \,|\, z \rangle = 0$. Consequently, $f(y) = 0_V$ for all $y \in V$ and therefore $f = 0$. $\square$

Theorem 8.12

Let V be a complex inner product space and let $f : V \to V$ be linear. Then the following statements are equivalent:

(1) *f is self-adjoint;*
(2) *$(\forall x \in V) \quad \langle f(x) \,|\, x \rangle \in \mathbb{R}$.*

Proof

$(1) \Rightarrow (2)$: If f is self-adjoint then for every $x \in V$ we have

$$\overline{\langle f(x) \,|\, x \rangle} = \overline{\langle f^*(x) \,|\, x \rangle} = \langle x \,|\, f^*(x) \rangle = \langle f(x) \,|\, x \rangle,$$

from which (2) follows.

$(2) \Rightarrow (1)$: If (2) holds then for every $x \in V$ we have

$$\langle f^*(x) \,|\, x \rangle = \langle x \,|\, f(x) \rangle = \overline{\langle f(x) \,|\, x \rangle} = \langle f(x) \,|\, x \rangle$$

and consequently

$$\langle (f^* - f)(x) \,|\, x \rangle = \langle f^*(x) \,|\, x \rangle - \langle f(x) \,|\, x \rangle = 0.$$

Since this holds for all $x \in V$ it follows by Theorem 8.11 that $f^* = f$. $\square$

This characterisation now leads to the following notion.

Definition

If V is an inner product space then a linear mapping $f : V \to V$ is said to be **positive** if it is self-adjoint and such that $\langle f(x) \,|\, x \rangle \geq 0$ for every $x \in V$; and **positive definite** if it is self-adjoint and such that $\langle f(x) \,|\, x \rangle > 0$ for every non-zero $x \in V$.

Example 8.5

Consider the linear mapping $f : \mathbb{R}^2 \to \mathbb{R}^2$ whose matrix, relative to the standard ordered basis, is the rotation matrix

$$A = \begin{bmatrix} \cos \vartheta & \sin \vartheta \\ -\sin \vartheta & \cos \vartheta \end{bmatrix},$$

i.e. f is given by $f(x,y) = (x \cos \vartheta + y \sin \vartheta, -x \sin \vartheta + y \cos \vartheta)$. For f to be positive it is necessary that A be symmetric, so that $\sin \vartheta = 0$; i.e. $\vartheta = n\pi$ for some integer n. Also, we must have

$$\langle f(x,y) \,|\, (x,y) \rangle = (x^2 + y^2) \cos \vartheta \geqslant 0.$$

This requires $(4k - 1)\frac{\pi}{2} \leqslant \vartheta \leqslant (4k + 1)\frac{\pi}{2}$. Consequently we see that f is positive if and only if $\vartheta = 2n\pi$ for some integer n, i.e. if and only if $f = \mathrm{id}$.

Theorem 8.13

If V is a non-zero finite-dimensional inner product space and $f : V \to V$ is linear then the following statements are equivalent:

(1) *f is positive;*
(2) *f is self-adjoint and every eigenvalue is real and greater than or equal to 0;*
(3) *there is a self-adjoint $g \in \mathrm{Lin}(V,V)$ such that $g^2 = f$;*
(4) *there exists $h \in \mathrm{Lin}(V,V)$ such that $h^\star \circ h = f$.*

Proof

(1) $\Rightarrow$ (2) : Let λ be an eigenvalue of f. By Theorem 8.9, λ is real. Then $0 \leqslant \langle f(x) \,|\, x \rangle = \langle \lambda x \,|\, x \rangle = \lambda \langle x \,|\, x \rangle$ gives $\lambda \geqslant 0$ since $\langle x \,|\, x \rangle > 0$.

(2) $\Rightarrow$ (3) : Since f is self-adjoint it is normal and hence is ortho-diagonalisable. Let its spectral resolution be $f = \sum\limits_{i=1}^{k} \lambda_i p_i$ and define $g : V \to V$ by

$$g = \sum_{i=1}^{k} \sqrt{\lambda_i} p_i.$$

Since the p_i are ortho-projections, and hence are self-adjoint, we see from (2) that g is self-adjoint. Also, since $p_i \circ p_j = 0$ when $i \neq j$, it follows readily that $g^2 = f$.

(3) $\Rightarrow$ (4) : Take $h = g$.

(4) $\Rightarrow$ (1) : Observe that $(h^\star \circ h)^\star = h^\star \circ h^{\star\star} = h^\star \circ h$ and that, for all $x \in V$, $\langle h^\star[h(x)] \,|\, x \rangle = \langle h(x) \,|\, h(x) \rangle \geqslant 0$. Thus we see that $f = h^\star \circ h$ is positive. $\square$

It is immediate from Theorem 8.13 that every positive linear mapping has a square root. We shall now show in fact that f has a unique positive square root.

Theorem 8.14

Let f be a positive linear mapping on a non-zero finite-dimensional inner product space V. Then there is a unique positive linear mapping $g : V \to V$ such that $g^2 = f$. Moreover, $g = q(f)$ for some polynomial q.

Proof

Let $f = \sum\limits_{i=1}^{k} \lambda_i p_i$ be the spectral resolution of f and define g as before, namely

$$g = \sum_{i=1}^{k} \sqrt{\lambda_i} p_i.$$

Since this must be the spectral resolution of g, it follows that the eigenvalues of g are $\sqrt{\lambda_i}$ for $i = 1, \dots, k$. It follows by Theorem 8.13 that g is positive.

Suppose now that $h : V \to V$ is also positive and such that $h^2 = f$. If the spectral resolution of h is $\sum\limits_{j=1}^{m} \mu_j q_j$ where the q_j are orthogonal projections then we have

$$\sum_{i=1}^{k} \lambda_i p_i = f = h^2 = \sum_{j=1}^{m} \mu_j^2 q_j.$$

Now, as in the proof of Theorem 8.5, the eigenspaces of f are the subspaces $\operatorname{Im} p_i$ for $i = 1, \dots, k$, and also $\operatorname{Im} q_j$ for $j = 1, \dots m$. It follows that $m = k$ and that there is a permutation σ on $\{1, \dots, k\}$ such that $q_{\sigma(i)} = p_i$, whence we have that $\mu_{\sigma(i)}^2 = \lambda_i$. Thus $\mu_{\sigma(i)} = \sqrt{\lambda_i}$ and we deduce that $h = g$.

For the final statement, consider the Lagrange polynomials

$$L_i = \prod_{j \ne i} \frac{X - \lambda_j}{\lambda_i - \lambda_j}.$$

Since $L_i(\lambda_t) = \delta_{it}$, the polynomial $q = \sum\limits_{i=1}^{k} \sqrt{\lambda_i} L_i$ is then such that $q(f) = g$. □

There is a corresponding result to Theorem 8.13 for positive definite linear mappings, namely:

Theorem 8.15

If V is a non-zero finite-dimensional inner product space and if $f : V \to V$ is linear then the following statements are equivalent:

(1) *f is positive definite;*
(2) *f is self-adjoint and all the eigenvalues of f are real and strictly positive;*
(3) *there is an invertible self-adjoint g such that $g^2 = f$;*
(4) *there is an invertible h such that $h^* \circ h = f$.*

Proof

This follows immediately from Theorem 8.13 on noting that g is invertible if and only if 0 is not one of its eigenvalues. □

Corollary

If f is positive definite then f is invertible. □

Of course, the above results have matrix analogues. A matrix that represents a positive (definite) linear mapping with respect to some ordered orthonormal basis in a finite-dimensional inner product space is called a **positive (definite) matrix**. A positive matrix is often also called a **Gram matrix**. By Theorem 8.13, we have the following characterisation.

Theorem 8.16

For a square matrix A the following statements are equivalent:

(1) *A is a Gram matrix;*
(2) *A is self-adjoint and all its eigenvalues are greater than or equal to 0;*
(3) *there is a self-adjoint matrix B such that $B^2 = A$;*
(4) *there is a matrix C such that $C^*C = A$.*   □

Corollary

A real symmetric matrix is positive definite if and only if all its eigenvalues are real and strictly positive.   □

EXERCISES

8.19 Let $f : \mathbb{R}^3 \rightarrow \mathbb{R}^3$ be the linear mapping that is represented, relative to the standard ordered basis, by the matrix

$$A = \begin{bmatrix} \frac{3}{2} & \frac{1}{2} & -1 \\ \frac{1}{2} & \frac{3}{2} & -1 \\ -1 & -1 & 3 \end{bmatrix}.$$

Prove that f is positive definite.

8.20 If V is a finite-dimensional inner product space and if $f, g : V \rightarrow V$ are positive definite, prove that so also is $f + g$.

8.21 If V is a finite-dimensional inner product space and if $f, g : V \rightarrow V$ are positive definite, prove that $f \circ g$ is positive definite if and only if f and g commute.

8.22 Let V be a finite-dimensional vector space with a given inner product $\langle - | - \rangle$. Prove that every inner product on V can be expressed in the form $\langle f(-) | - \rangle$ for a unique positive definite $f \in \mathrm{Lin}\,(V, V)$.

[*Hint.* Let $\langle\!\langle - | - \rangle\!\rangle$ be an inner product on V. Use Exercise 7.23 to produce, for the linear form $\langle\!\langle - | \beta \rangle\!\rangle \in V^d$, a unique element $\beta' \in V$ such that $\langle\!\langle - | \beta \rangle\!\rangle = \langle - | \beta' \rangle$. Define $f : V \rightarrow V$ by $f(\beta) = \beta'$.]

9
Bilinear and Quadratic Forms

In this chapter we shall apply some of the previous results in a study of certain types of linear forms.

Definition

Let V and W be vector spaces over a field F. A **bilinear form** on $V \times W$ is a mapping $f : V \times W \to F$ such that, for all $x \in V$, all $y \in W$, and all $\lambda \in F$,

(1) $f(x + x', y) = f(x, y) + f(x', y)$;

(2) $f(x, y + y') = f(x, y) + f(x, y')$;

(3) $f(\lambda x, y) = \lambda f(x, y) = f(x, \lambda y)$.

Example 9.1

The standard inner product on $\mathbb{R}^n$, namely $f\big((x_1, \ldots, x_n), (y_1, \ldots, y_n)\big) = \sum_{i=1}^{n} x_i y_i$ is a bilinear form on $\mathbb{R}^n \times \mathbb{R}^n$.

Example 9.2

Let V be the vector space of real continuous functions. Let $K(x, y)$ be a given continuous function of two real variables and let $a, b \in \mathbb{R}$. Define $H : V \times V \to \mathbb{R}$ by

$$H(g, h) = \int_a^b \int_a^b K(x, y) g(x) h(y) \, dx \, dy.$$

Then standard properties of integrals show that H is bilinear.

EXERCISE

9.1 Determine if $f : \mathbb{R}^2 \times \mathbb{R}^2 \to \mathbb{R}$ is bilinear when $f\big((x_1, y_1), (x_2, y_2)\big)$ is

(1) $x_1 y_2 - x_2 y_1$;

(2) $(x_1 + y_1)^2 - x_2 y_2$.

Definition

Let V be a vector space of dimension n over F and let $(v_i)_n$ be an ordered basis of V. If $f : V \times V \to F$ is a bilinear form then by the **matrix of f relative to the ordered basis** $(v_i)_n$ we shall mean the matrix $A = [a_{ij}]_{n \times n}$ given by $a_{ij} = f(v_i, v_j)$.

Suppose now that V is of dimension n and that $f : V \times V \to F$ is bilinear. Relative to an ordered basis $(v_i)_n$ let $x = \sum_{i=1}^{n} x_i v_i$ and $y = \sum_{i=1}^{n} y_i v_i$. If the matrix of f relative to $(v_i)_n$ is $A = [a_{ij}]$ then, by the bilinearity of f, we see that

$$(\star) \qquad f(x, y) = \sum_{i=1}^{n} \sum_{j=1}^{n} x_i y_j f(v_i, v_j) = \sum_{i,j=1}^{n} x_i y_j a_{ij}.$$

Conversely, given any $n \times n$ matrix $A = [a_{ij}]$ over F it is easy to see that $(\star)$ defines a bilinear form f on $V \times V$; simply observe that the above can be written as

$$f(x, y) = x^t A y = \begin{bmatrix} x_1 & \cdots & x_n \end{bmatrix} A \begin{bmatrix} y_1 \\ \vdots \\ y_n \end{bmatrix}.$$

Moreover, with respect to the ordered basis $(v_i)_n$ we have

$$f(v_i, v_j) = \begin{bmatrix} 0 & \cdots & 1_{(i)} & \cdots & 0 \end{bmatrix} A \begin{bmatrix} 0 \\ \vdots \\ 1_{(j)} \\ \vdots \\ 0 \end{bmatrix}$$

$$= a_{ij},$$

and so the matrix of f is A.

Example 9.3

In Example 9.1 the matrix of f relative to the standard ordered basis is I_n.

Example 9.4

The matrix

$$A = \begin{bmatrix} 1 & 2 & 5 \\ -2 & 0 & 1 \\ 0 & -6 & 6 \end{bmatrix}$$

gives rise to the bilinear form

$$x^t A y = x_1 y_1 + 2x_1 y_2 + 5x_1 y_3 - 2x_2 y_1 + x_2 y_3 - 6x_3 y_2 + 6x_3 y_3.$$

Example 9.5

The matrix

$$A = \begin{bmatrix} a & h & g \\ h & b & f \\ g & f & c \end{bmatrix}$$

gives rise to the bilinear form

$$x^t A y = a x_1 y_1 + b x_2 y_2 + c x_3 y_3 + h(x_1 y_2 + x_2 y_1) + g(x_1 y_3 + x_3 y_1) + f(x_2 y_3 + x_3 y_2).$$

Example 9.6

The bilinear form $x_1(y_2 + y_3) + x_2 y_3$ is represented, relative to the standard ordered basis, by the matrix

$$A = \begin{bmatrix} 0 & 1 & 1 \\ 0 & 0 & 1 \\ 0 & 0 & 0 \end{bmatrix}.$$

EXERCISE

9.2 Determine the matrix of each of the bilinear forms

(1) $2x_1 y_1 - 3x_1 y_3 + 2x_2 y_2 - 5x_2 y_3 + 4x_3 y_1$;

(2) $3x_1 y_1 + 2x_2 y_2 + x_3 y_3$.

It is natural to ask how a change of basis affects the matrix of a bilinear form.

Theorem 9.1

Let V be a vector space of dimension n over a field F. Let $(v_i)_n$ and $(w_i)_n$ be ordered bases of V. If $f : V \times V \to F$ is bilinear and if A is the matrix of f relative to $(v_i)_n$ then the matrix of f relative to $(w_i)_n$ is $P^t A P$ where P is the transition matrix from $(w_i)_n$ to $(v_i)_n$.

Proof

We have $w_j = \sum_{i=1}^{n} p_{ij} v_i$ for $j = 1, \ldots, n$ and so, by the bilinearity of f,

$$f(w_i, w_j) = f\left(\sum_{t=1}^{n} p_{ti} v_t, \sum_{k=1}^{n} p_{kj} v_k \right)$$

$$= \sum_{t=1}^{n} \sum_{k=1}^{n} p_{ti} p_{kj} f(v_t, v_k)$$

$$= \sum_{t=1}^{n} \sum_{k=1}^{n} p_{ti} p_{kj} a_{tk}$$

$$= [P^t A P]_{ij},$$

from which the result follows. $\square$

Example 9.7

Consider the bilinear form $2x_1y_1 - 3x_2y_2 + x_3y_3$ relative to the standard ordered basis of $\mathbb{R}^3$. The matrix that represents this is

$$A = \begin{bmatrix} 2 & 0 & 0 \\ 0 & 0 & -3 \\ 0 & 0 & 1 \end{bmatrix}.$$

To compute the form resulting in a change of reference to the ordered basis

$$\{(1, 1, 1), (-2, 1, 1), (2, 1, 0)\},$$

we observe that the transition matrix from this new basis to the old basis is

$$P = \begin{bmatrix} 1 & -2 & 2 \\ 1 & 1 & 1 \\ 1 & 1 & 0 \end{bmatrix}.$$

Consequently the matrix of the form that results from the change of basis is

$$P^tAP = \begin{bmatrix} 0 & -6 & 4 \\ -6 & 6 & -8 \\ 1 & -11 & 8 \end{bmatrix}.$$

The form relative to the new basis can be read off from this matrix.

EXERCISE

9.3 Consider the bilinear form $f : \mathbb{R}^3 \to \mathbb{R}^3$ given with reference to the standard ordered basis of $\mathbb{R}^3$ by

$$2x_1y_1 + x_2y_3 - 3x_3y_2 + x_3y_3.$$

Determine its equivalent form with reference to the basis

$$\{(1, 2, 3), (-1, 1, 2), (1, 2, 1)\}.$$

Definition

If A and B are $n \times n$ matrices over a field F then we say that B is **congruent** to A if there is an invertible matrix P such that $B = P^tAP$.

It is readily seen that the relation of being congruent is an equivalence relation on $\text{Mat}_{n \times n} F$.

EXERCISE

9.4 Prove that any matrix that is congruent to a symmetric matrix is also symmetric.

Definition

A bilinear form $f : V \times V \to V$ is said to be **symmetric** if $f(x, y) = f(y, x)$ for all $x, y \in V$; and **skew-symmetric** if $f(x, y) = -f(y, x)$ for all $x, y \in V$.

EXERCISE

9.5 Prove that every bilinear form can be expressed uniquely as the sum of a symmetric and a skew-symmetric bilinear form. Do so for

(1) $x_1 y_1 + x_1 y_2 + 2x_2 y_1 + x_2 y_2$;

(2) $2x_1 y_1 - 3x_2 y_2 + x_3 y_3$.

It is clear that a matrix that represents a symmetric bilinear form is symmetric; and, conversely, that every symmetric matrix gives rise to a symmetric bilinear form.

Definition

If V is a vector space over a field F then by a **quadratic form** on V we mean a mapping $Q : V \to F$ given, for some symmetric bilinear form $f : V \times V \to F$, by

$$(\forall x \in V) \qquad Q(x) = f(x, x).$$

Example 9.8

The mapping $Q : \mathbb{R}^2 \to \mathbb{R}^2$ given by

$$Q(x, y) = x^2 - xy + y^2$$

is a quadratic form on $\mathbb{R}^2$, as can be seen on writing the right hand side as

$$x^t A x = \begin{bmatrix} x & y \end{bmatrix} \begin{bmatrix} 1 & -\frac{1}{2} \\ -\frac{1}{2} & 1 \end{bmatrix} \begin{bmatrix} x \\ y \end{bmatrix}.$$

Example 9.9

The mapping $Q : \mathbb{R}^3 \to \mathbb{R}^3$ given by

$$Q(x, y, z) = x^2 + y^2 - z^2$$

is a quadratic form on $\mathbb{R}^3$, as can be seen on writing the right hand side as

$$x^t A x = \begin{bmatrix} x & y & z \end{bmatrix} \begin{bmatrix} 1 & 0 & 0 \\ 0 & 1 & 0 \\ 0 & 0 & -1 \end{bmatrix} \begin{bmatrix} x \\ y \\ z \end{bmatrix}.$$

EXERCISES

9.6 Prove that the set of quadratic forms on a vector space V forms a subspace of the vector space $\text{Map}(V, F)$.

9.7 Relative to the symmetric bilinear form provided by the standard inner product on $\mathbb{R}^n$, describe the associated quadratic form.

9.8 Prove that the following expressions are quadratic forms:

(1) $3x^2 - 5xy - 7y^2$;

(2) $3x^2 - 7xy + 5xz + 4y^2 - 4yz - 3z^2$.

In what follows we shall restrict the ground field F to be $\mathbb{R}$.

Given a symmetric bilinear form $f : V \times V \to \mathbb{R}$, by the **associated quadratic form** we shall mean the mapping $Q_f : V \to \mathbb{R}$ defined by $Q_f(x) = f(x, x)$.

Theorem 9.2

Let V be a vector space over $\mathbb{R}$. If $f : V \times V \to \mathbb{R}$ is a symmetric bilinear form then the following identities hold:

(1) $Q_f(\lambda x) = \lambda^2 Q_f(x)$;

(2) $f(x, y) = \frac{1}{2}[Q_f(x + y) - Q_f(x) - Q_f(y)]$;

(3) $f(x, y) = \frac{1}{4}[Q_f(x + y) - Q_f(x - y)]$.

Proof

(1) $Q_f(\lambda x) = f(\lambda x, \lambda x) = \lambda^2 f(x, x) = \lambda^2 Q_f(x)$.

(2) Since f is symmetric, we have

$$Q_f(x + y) = f(x + y, x + y) = Q_f(x) + 2f(x, y) + Q_f(y).$$

(3) By (1) we have $Q_f(-x) = Q_f(x)$ and so, by (2),

$$Q_f(x - y) = Q_f(x) - 2f(x, y) + Q_f(y).$$

This, together with (2), gives (3). $\square$

An important consequence of the above is the following.

Theorem 9.3

Every real quadratic form is associated with a uniquely determined symmetric bilinear form.

Proof

Suppose that $Q : V \to \mathbb{R}$ is a quadratic form. Suppose further that $f, g : V \times V \to \mathbb{R}$ are symmetric bilinear forms such that $Q = Q_f = Q_g$. Then by Theorem 9.2 we see that $f(x, y) = g(x, y)$ for all $x, y \in V$ and therefore $f = g$. $\square$

Because of this fact, we define the **matrix of a real quadratic form** on a finite-dimensional vector space to be the matrix of the associated symmetric bilinear form.

Example 9.10

The mapping $Q : \mathbb{R}^2 \to \mathbb{R}$ given by

$$Q(x, y) = 4x^2 + 6xy + 9y^2 = \begin{bmatrix} x & y \end{bmatrix} \begin{bmatrix} 4 & 3 \\ 3 & 9 \end{bmatrix} \begin{bmatrix} x \\ y \end{bmatrix}$$

is a quadratic form. The matrix of Q is

$$\begin{bmatrix} 4 & 3 \\ 3 & 9 \end{bmatrix}$$

and the associated bilinear form is

$$f\big((x, y), (x', y')\big) = 4xx' + 3(xy' + x'y) + 9yy'.$$

EXERCISE

9.9 Given a quadratic form $Q : \mathbb{R}^n \to \mathbb{R}$, let A be the matrix of Q and let λ be an eigenvalue of A. Prove that there exist $a_1, \ldots, a_n$, not all zero, such that

$$Q(a_1, \ldots, a_n) = \lambda \sum_{i=1}^{n} a_i^2.$$

It is clear from Theorem 9.1 that symmetric matrices A, B represent the same quadratic form relative to different ordered bases if and only if they are congruent. Our objective now is to obtain a canonical form for real symmetric matrices under congruence, i.e. to obtain a particularly simple representative in each congruence class. This will then provide us with a canonical form for the associated real quadratic form. The results on orthogonal similarity that we have obtained in Chapter 8 will put us well on the road.

For our immediate purposes, we note that if A and B are matrices that are congruent then A and B have the same rank. In fact, $B = P^t A P$ for some invertible matrix P. Since an invertible matrix is a product of elementary matrices, multiplication by which can be expressed in terms of elementary row and column operations, all of which leave the rank invariant, the rank of $P^t A P$ is the same as the rank of A.

Theorem 9.4

If A is an $n \times n$ real symmetric matrix then A is congruent to a unique matrix of the form

$$\begin{bmatrix} I_r & & \\ & -I_s & \\ & & 0 \end{bmatrix}.$$

Proof

Since A is real symmetric it follows by the Corollary to Theorem 8.10 and the Corollary to Theorem 8.9 that A is orthogonally similar to a diagonal matrix and all the eigenvalues of A are real. Let the positive eigenvalues be $\lambda_1, \ldots, \lambda_r$ and let the negative eigenvalues be $-\lambda_{r+1}, \ldots, -\lambda_{r+s}$. Then there is an orthogonal matrix P such that

$$P^t A P = \mathrm{diag}\{\lambda_1, \ldots, \lambda_r, -\lambda_{r+1}, \ldots, -\lambda_{r+s}, 0, \ldots, 0\},$$

there being $n - (r + s)$ entries 0.

Let $D = [d_{ij}]_{n \times n}$ be the diagonal matrix such that

$$d_{ii} = \begin{cases} \frac{1}{\sqrt{\lambda_i}} & \text{if } i = 1, \ldots, r + s; \\ 1 & \text{otherwise.} \end{cases}$$

Then it is readily seen that

$$(PD)^t A P D = D P^t A P D = \begin{bmatrix} I_r & & \\ & -I_s & \\ & & 0 \end{bmatrix}.$$

Now since P and D are both invertible, so is PD. Thus we see that A is congruent to a matrix of the stated form.

As for uniqueness, it suffices to suppose that

$$L = \begin{bmatrix} I_r & & \\ & -I_s & \\ & & 0 \end{bmatrix}, \quad M = \begin{bmatrix} I_{r'} & & \\ & -I_{s'} & \\ & & 0 \end{bmatrix}$$

are congruent and show that $r = r'$ and $s = s'$.

Now if L and M are congruent then they have the same rank, so $r + s = r' + s'$. Suppose now, by way of obtaining a contradiction, that $r < r'$ (in which case $s' < s$). Let W be the real vector space $\mathrm{Mat}_{n \times 1}\,\mathbb{R}$. Clearly, W is an inner product space under the definition

$$\langle \mathbf{x} \,|\, \mathbf{y} \rangle = \mathbf{x}^t \mathbf{y}.$$

Consider the mapping $f_L : W \to W$ given by $f_L(\mathbf{x}) = L\mathbf{x}$. Since L is symmetric we have

$$\langle L\mathbf{x} \,|\, \mathbf{y} \rangle = (L\mathbf{x})^t \mathbf{y} = \mathbf{x}^t L\mathbf{y} = \langle \mathbf{x} \,|\, L\mathbf{y} \rangle$$

and so f_L is self-adjoint. Likewise, so also is the mapping $f_M : W \to W$ given by $f_M(\mathbf{x}) = M\mathbf{x}$. Consider now the subspaces

$$X = \{\mathbf{x} \in W ; \ x_1 = \cdots = x_r = 0, \ x_{r+s+1} = \cdots = x_n = 0\};$$

$$Y = \{\mathbf{x} \in W ; \ x_{r'+1} = \cdots = x_{r'+s'} = 0\}.$$

Clearly, X is of dimension s, and for every non-zero $\mathbf{x} \in X$ we have

$$(1) \qquad \mathbf{x}'L\mathbf{x} = -x_{r+1}^2 - \cdots - x_{r+s}^2 < 0.$$

Also, Y is of dimension $n - s'$, and for every $\mathbf{x} \in Y$ we have

$$\mathbf{x}'M\mathbf{x} = x_1^2 + \cdots + x_{r'}^2 \geqslant 0.$$

Now since L and M are congruent there is an invertible matrix P such that $M = P'LP$. Defining $f_P : W \to W$ by $f_P(\mathbf{x}) = P\mathbf{x}$ and observing that f_P is also self-adjoint, we then have, for all $\mathbf{x} \in Y$,

$$0 \leqslant \mathbf{x}'M\mathbf{x} = \langle M\mathbf{x}\,|\,\mathbf{x}\rangle = \langle P'LP\mathbf{x}\,|\,\mathbf{x}\rangle = \langle (f_L \circ f_P)(\mathbf{x})\,|\,f_P(\mathbf{x})\rangle,$$

from which we see that if $Z = \{f_P(\mathbf{x}) \; ; \; \mathbf{x} \in Y\}$ then

$$(2) \qquad (\forall \mathbf{z} \in Z) \quad \langle f_L(\mathbf{z})\,|\,\mathbf{z}\rangle \geqslant 0.$$

Now since f_P is an isomorphism we have $\dim Z = \dim Y = n - s'$ and so

$$\dim Z + \dim X = n - s' + s > n \geqslant \dim(Z + X).$$

It follows that the sum $Z + X$ is not direct (for otherwise we would have equality), and so $Z \cap X \neq \{0_W\}$. Consider now a non-zero element $\mathbf{z}$ of $Z \cap X$. From (1) we see that $\langle f_L(\mathbf{z})\,|\,\mathbf{z}\rangle$ is negative, whereas from (2) we see that $\langle f_L(\mathbf{z})\,|\,\mathbf{z}\rangle$ is non-negative. This contradiction shows that we cannot have $r' < r$. Similarly, we cannot have $r < r'$. We therefore conclude that $r = r'$ whence also $s = s'$. $\square$

The above result gives immediately the following theorem which describes the canonical quadratic forms.

Theorem 9.5

[Sylvester]   *Let V be a vector space of dimension n over $\mathbb{R}$ and let $Q : V \to \mathbb{R}$ be a quadratic form on V. Then there is an ordered basis $(v_i)_n$ of V such that if $x = \sum_{i=1}^{n} x_i v_i$ then*

$$Q(x) = x_1^2 + \cdots + x_r^2 - x_{r+1}^2 - \cdots - x_{r+s}^2.$$

Moreover, the integers r and s are independent of such a basis. $\square$

Definition

The integer $r + s$ in Theorem 9.5 is called the **rank** of the quadratic form Q, and the integer $r - s$ is called the **signature** of Q.

Example 9.11

Consider the quadratic form $Q : \mathbb{R}^3 \to \mathbb{R}$ given by

$$Q(x, y, z) = x^2 - 2xy + 4yz - 2y^2 + 4z^2.$$

The matrix of Q is the symmetric matrix

$$A = \begin{bmatrix} 1 & -1 & 0 \\ -1 & -2 & 2 \\ 0 & 2 & 4 \end{bmatrix}.$$

We can use the method outlined in the proof of Theorem 9.4 to effect a matrix reduction and thereby obtain the canonical form of Q.

In many cases, however, it is often easier to avoid such calculations by judicious use of the method of 'completing the squares'. For example, for the above Q it is readily seen that

$$Q(x, y, z) = (x - y)^2 + (y + 2z)^2 - 4y^2$$

which is in canonical form. Thus Q is of rank 3 and of signature 1.

Example 9.12

The quadratic form $Q : \mathbb{R}^3 \rightarrow \mathbb{R}$ given by $Q(x, y, z) = 2xy + 2yz$ can be reduced to canonical form either by the method of completing squares or by a matrix reduction. The former method is not so easy in this case, but can be achieved as follows. Define

$$\sqrt{2}x = X + Y, \quad \sqrt{2}y = X - Y, \quad \sqrt{2}z = Z.$$

Then

$$\begin{aligned} Q(x, y, z) &= (X + Y)(X - Y) + (X - Y)Z \\ &= X^2 - Y^2 + (X - Y)Z \\ &= (X + \tfrac{1}{2}Z)^2 - (Y + \tfrac{1}{2}Z)^2 \\ &= \tfrac{1}{2}(x + y + z)^2 - \tfrac{1}{2}(x - y + z)^2, \end{aligned}$$

which is of rank 2 and signature 0.

Example 9.13

Consider the quadratic form $Q : \mathbb{R}^3 \rightarrow \mathbb{R}$ given by

$$Q(x, y, z) = 4xy + 2yz.$$

Here we have

$$\begin{aligned} 4xy + 2yz &= (x + y)^2 - (x - y)^2 + 2yz \\ &= X^2 - Y^2 + (X - Y)z \qquad [X = x + y, \ Y = x - y] \\ &= (X + \tfrac{1}{2}z)^2 - Y^2 - Yz - \tfrac{1}{4}z^2 \\ &= (X + \tfrac{1}{2}z)^2 - (Y + \tfrac{1}{2}z)^2 \\ &= \xi^2 - \eta^2, \end{aligned}$$

where $\xi = x + y + \frac{1}{2}z$ and $\eta = x - y + \frac{1}{2}z$. Taking $\zeta = z$ we then have

$$x = \tfrac{1}{2}(\xi + \eta - \zeta)$$
$$y = \tfrac{1}{2}(\xi - \eta)$$
$$z = \zeta,$$

and so if we define

$$P = \begin{bmatrix} \frac{1}{2} & \frac{1}{2} & -\frac{1}{2} \\ \frac{1}{2} & -\frac{1}{2} & 0 \\ 0 & 0 & 1 \end{bmatrix}$$

then we have

$$\begin{bmatrix} x \\ y \\ z \end{bmatrix} = P \begin{bmatrix} \xi \\ \eta \\ \zeta \end{bmatrix}.$$

To see that P does the reduction, observe that the matrix of Q is

$$A = \begin{bmatrix} 0 & 2 & 0 \\ 2 & 0 & 1 \\ 0 & 1 & 0 \end{bmatrix}.$$

A simple check gives $P^t A P = \text{diag}\{1, -1, 0\}$.

EXERCISES

9.10 A quadratic form $Q : \mathbb{R}^3 \to \mathbb{R}$ is represented, relative to the standard ordered basis of $\mathbb{R}^3$ by the matrix

$$A = \begin{bmatrix} 1 & 1 & -1 \\ 1 & 1 & 0 \\ -1 & 0 & -1 \end{bmatrix}.$$

Determine the matrix of the canonical form of Q.

9.11 By completing squares, determine the rank and the signature of each of the quadratic forms

(1) $2y^2 - z^2 + xy + xz$;

(2) $2xy - xz - yz$;

(3) $yz + xz + xy + xt + yt + zt$.

9.12 For each of the following quadratic forms determine the canonical form, and an invertible matrix that does the reduction:

(1) $x^2 + 2y^2 + 9z^2 - 2xy + 4xz - 6yz$;

(2) $x^2 + 4y^2 + z^2 - 4t^2 + 2xy - 2xt + 6yz - 8yt - 14zt$.

Definition

A quadratic form Q is said to be **positive definite** if $Q(x) > 0$ for all $x \neq 0_V$.

If Q is a quadratic form on a finite-dimensional vector space and if A is its matrix then Q is positive definite if and only if

$$0 < Q(x) = x^t A x = \langle Ax \mid x \rangle,$$

which is the case if and only if the corresponding symmetric bilinear form is positive definite or, equivalently, the matrix A is positive definite.

Theorem 9.6

A real quadratic form on a finite-dimensional vector space is positive definite if and only if its rank and signature are the same.

Proof

Clearly, such a form is positive definite if and only if there are no negative terms in its canonical form. This is the case if and only if the rank and the signature are the same. $\square$

Example 9.14

Let $f : \mathbb{R} \times \mathbb{R} \to \mathbb{R}$ be a function whose partial derivatives f_x, f_y are zero at the point (x_0, y_0). Then the Taylor series at $(x_0 + h, y_0 + h)$ is

$$f(x_0, y_0) + \tfrac{1}{2}[h^2 f_{xx} + 2hk f_{xy} + k^2 f_{yy}](x_0, y_0) + \cdots .$$

For small values of h, k the significant term in this is the quadratic form

$$[h^2 f_{xx} + 2hk f_{xy} + k^2 f_{yy}](x_0, y_0)$$

in h, k. If it has rank 2 then its canonical form is $\pm H^2 \pm K^2$. If both signs are positive (i.e. the form is positive definite) then f has a relative minimum at (x_0, y_0), and if both signs are negative then f has a relative maximum at (x_0, y_0). If one sign is positive and the other is negative then f has a saddle point at (x_0, y_0). Thus the geometry is distinguished by the signature of the quadratic form.

EXERCISES

9.13 Consider the quadratic form $Q : \mathbb{R}^3 \to \mathbb{R}$ that is represented, relative to the standard basis, by the matrix

$$A = \begin{bmatrix} 1 & 1 & -1 \\ 1 & 1 & 0 \\ -1 & 0 & -1 \end{bmatrix}.$$

Is Q positive definite?

9.14 Show that neither of the quadratic forms defined on $\mathbb{R}^2$ by

(1) $Q(x, y) = x^2 + 3xy + y^2$;

(2) $Q(x, y) = 2x^2 - 4xy + 3y^2 - z^2$,

is positive definite.

9.15 Show that for the quadratic form Q defined on $\mathbb{R}^n$ by

$$Q(x_1, \ldots, x_n) = \sum_{r<s}(x_r - x_s)^2$$

both the rank and the signature are $n - 1$.

[*Hint.* Show that the matrix of Q relative to the standard ordered basis is

$$A = \begin{bmatrix} n-1 & -1 & -1 & \cdots & -1 \\ -1 & n-1 & -1 & \cdots & -1 \\ -1 & -1 & n-1 & \cdots & -1 \\ \vdots & \vdots & \vdots & \ddots & \vdots \\ -1 & -1 & -1 & \cdots & n-1 \end{bmatrix}.$$

Show that A is congruent to $\mathrm{diag}\{n-1, n-2, \ldots, 2, 1, 0\}$.]

9.16 Consider the quadratic form

$$\sum_{r,s=1}^{n}(\lambda rs + r + s)x_r x_s$$

in which $\lambda \in \mathbb{R}$ is arbitrary. Show that with $y_1 = \sum_{r=1}^{n} rx_r$ and $y_2 = \sum_{i=1}^{n} x_r$ the form becomes

$$\lambda y_1^2 + 2y_1 y_2.$$

Hence show that for all values of λ the rank is 2 and the signature is 0.

We have seen in Theorem 8.8 that the ortho-diagonalisable linear mappings on a complex inner product space are precisely those that are normal; and in Theorem 8.10 that the ortho-diagonalisable linear mappings on a real inner product space are precisely those that are self-adjoint. It is therefore natural to ask what can be said about *normal* linear mappings on a *real* inner product space; equivalently, we may ask about real square matrices that commute with their transposes. Our objective here is to obtain a canonical form for such a matrix under orthogonal similarity.

As we shall see, the main results that we shall obtain stem from further applications of the Primary Decomposition Theorem. The notion of minimum polynomial will therefore play an important part in this. Now, as we are assuming throughout that the ground field is $\mathbb{R}$, it is clear that we shall require a knowledge of what the monic irreducible polynomials over $\mathbb{R}$ look like. This is the content of the following result.

Theorem 10.1

A monic polynomial f over $\mathbb{R}$ is irreducible if and only if it is of the form $X - a$ for some $a \in \mathbb{R}$, or of the form $(X - a)^2 + b^2$ for some $a, b \in \mathbb{R}$ with $b \neq 0$.

Proof

It is clear that $X - a$ is irreducible over $\mathbb{R}$. Consider now the polynomial

$$f = (X - a)^2 + b^2 = X^2 - 2aX + a^2 + b^2$$

where $a, b \in \mathbb{R}$ with $b \neq 0$. By way of obtaining a contradiction, suppose that f is reducible over $\mathbb{R}$. Then there exist $p, q \in \mathbb{R}$ such that

$$f = (X - p)(X - q) = X^2 - (p + q)X + pq.$$

It follows that $p + q = 2a$ and $pq = a^2 + b^2$. These equations give

$$p^2 - 2ap + a^2 + b^2 = 0$$

and so we have that

$$p = \tfrac{1}{2}[2a \pm \sqrt{4a^2 - 4(a^2 + b^2)}] = a \pm \sqrt{-b^2}.$$

Since $b \neq 0$ by hypothesis, we have the contradiction $p \notin \mathbb{R}$.

Conversely, suppose that $f \in \mathbb{R}[X]$ is monic and irreducible. Suppose further that f is not of the form $X - a$ where $a \in \mathbb{R}$. Let $z = a + ib \in \mathbb{C}$ be a root of f, noting that this exists by the fundamental theorem of algebra. If $f = \sum_{i=0}^{n} a_i X^i$ then

$0 = f(z) = \sum_{i=0}^{n} a_i z^i$. Taking complex conjugates, we obtain

$$0 = \overline{\sum_{i=0}^{n} a_i z^i} = \sum_{i=0}^{n} a_i \overline{z}^i = f(\overline{z}).$$

Thus we see that $\overline{z} \in \mathbb{C}$ is also a root of f. Consequently,

$$(X - z)(X - \overline{z}) = X^2 - (z + \overline{z})X + z\overline{z} = X^2 - 2aX + a^2 + b^2 \in \mathbb{R}[X]$$

is a divisor of f. Since by hypothesis f is irreducible over $\mathbb{R}$, we must have $z \in \mathbb{C} \backslash \mathbb{R}$ whence $b \neq 0$, and then $f = X^2 - 2aX + a^2 + b^2 = (X - a)^2 + b^2$. $\square$

Corollary

The irreducibles of $\mathbb{R}[X]$ are the polynomials of degree 1, *and the polynomials* $pX^2 + qX + r$ *where* $q^2 - 4pr < 0$.

Proof

If $pX^2 + qX + r$ were reducible over $\mathbb{R}$ then it would have a linear factor $X - \alpha$ with $\alpha \in \mathbb{R}$, whence $p\alpha^2 + q\alpha + r = 0$. For this to hold, it is necessary that $q^2 - 4pr \geqslant 0$. Consequently, if $q^2 - 4pr < 0$ then the polynomial $pX^2 + qX + r$ is irreducible.

Conversely, if $pX^2 + qX + r$ is irreducible over $\mathbb{R}$ then so is

$$X^2 + \frac{q}{p}X + \frac{r}{p}.$$

It follows by Theorem 10.1 that there exist $a, b \in \mathbb{R}$ with $b \neq 0$ such that

$$\frac{q}{p} = -2a, \quad \frac{r}{p} = a^2 + b^2.$$

Consequently,

$$q^2 - 4pr = 4a^2 p^2 - 4p^2(a^2 + b^2) = -4p^2 b^2 < 0$$

as required. $\square$

Example 10.1

The zeros of $X^{2p} - 1 \in \mathbb{R}[X]$ are $e^{k\pi i/p} = \cos \frac{k\pi}{p} + i \sin \frac{k\pi}{p}$ for $k = 0, 1, \ldots, 2p - 1$. As a product of irreducibles over $\mathbb{R}$,

$$X^{2p} - 1 = (X - 1)(X + 1) \prod_{k=1}^{p-1} \left(X^2 - 2X \cos \frac{k\pi}{p} + 1\right).$$

If V is a real inner product space then a linear mapping $f : V \to V$ is said to be **skew-adjoint** if $f^* = -f$. Correspondingly, a real square matrix A is said to be **skew-symmetric** if $A' = -A$.

Theorem 10.2

Let V be a non-zero finite-dimensional real inner product space and let $f : V \to V$ be linear. Then there is a unique self-adjoint mapping $g : V \to V$ and a unique skew-adjoint mapping $h : V \to V$ such that $f = g + h$. Moreover, f is normal if and only if g and h commute.

Proof

Clearly, we have the decomposition

$$f = \tfrac{1}{2}(f + f^*) + \tfrac{1}{2}(f - f^*)$$

in which $\tfrac{1}{2}(f + f^*)$ is self-adjoint, and $\tfrac{1}{2}(f - f^*)$ is skew-adjoint.

Suppose now that $f = g + h$ where g is self-adjoint and h is skew-adjoint. Then we have $f^* = g^* + h^* = g - h$. It follows from these equations that $g = \tfrac{1}{2}(f + f^*)$ and $h = \tfrac{1}{2}(f - f^*)$, which establishes the uniqueness of such a decomposition.

If now f is normal then $f \circ f^* = f^* \circ f$ gives

$$(g + h) \circ (g - h) = (g - h) \circ (g + h)$$

which reduces to $g \circ h = h \circ g$. Conversely, if g and h commute then it is readily seen that $f \circ f^* = g^2 - h^2 = f^* \circ f$ whence f is normal. $\square$

The reader will be familiar with the matrix equivalent of Theorem 10.2, that a real square matrix A can be expressed uniquely as the sum of a symmetric and a skew-symmetric matrix, namely $A = \tfrac{1}{2}(A + A') + \tfrac{1}{2}(A - A')$.

A characterisation of skew-adjoint linear mappings on a real inner product space is the following.

Theorem 10.3

If V is a non-zero finite-dimensional real inner product space then a linear mapping $f : V \to V$ is skew-adjoint if and only if

$$(\forall x \in V) \qquad \langle f(x) \mid x \rangle = 0.$$

Proof

$\Rightarrow$: If f is skew-adjoint then for every $x \in V$ we have

$$\langle f(x) \mid x \rangle = \langle x \mid -f(x) \rangle = -\langle x \mid f(x) \rangle = -\langle f(x) \mid x \rangle$$

and therefore $\langle f(x) \mid x \rangle = 0$.

$\Leftarrow$: If the condition holds then for all $x, y \in V$ we have

$$0 = \langle f(x + y) \mid x + y \rangle = \langle f(x) \mid y \rangle + \langle f(y) \mid x \rangle$$

which gives

$$\langle f(x) | y \rangle = -\langle f(y) | x \rangle = -\langle x | f(y) \rangle = \langle x | -f(y) \rangle.$$

It now follows by the uniqueness of adjoints that $f^* = -f$ and consequently f is skew-adjoint. $\square$

EXERCISES

10.1 If V is a non-zero finite-dimensional real inner product space prove that for every linear mapping $f : V \to V$ there is a unique self-adjoint linear mapping $g : V \to V$ such that

$$(\forall x \in V) \qquad \langle f(x) | x \rangle = \langle g(x) | x \rangle.$$

10.2 If A is a real skew-symmetric matrix and if g is a polynomial such that $g(A) = 0$ prove that $g(-A) = 0$. Deduce that the minimum polynomial of A contains no terms of odd degree.

We now make the following observation.

Theorem 10.4

If V is a non-zero finite-dimensional real inner product space and if $f : V \to V$ is normal then its minimum polynomial is of the form $m_f = \prod_{i=1}^{k} p_i$ where $p_1, \ldots, p_k$ are distinct irreducible polynomials.

Proof

We know that the minimum polynomial of f is of the general form $m_f = \prod_{i=1}^{k} p_i^{e_i}$ where $p_1, \ldots, p_k$ are distinct irreducibles. What we have to show here is that when f is normal every $e_i = 1$.

Suppose, by way of obtaining a contradiction, that we have $e_i \geqslant 2$ for a particular index i. If we write as usual $V_i = \mathrm{Ker}\, p_i(f)^{e_i}$ then we have, for every $x \in V_i$,

$$p_i(f)^{e_i - 1}(x) \in \mathrm{Im}\, p_i(f) \cap \mathrm{Ker}\, p_i(f).$$

But since f is normal so also is $p_i(f)$, by Theorem 8.6(2). It then follows by Theorem 8.6(3) that the restriction of $p_i(f)^{e_i-1}$ to V_i is the zero mapping. If as usual we let $f_i : V_i \to V_i$ be the mapping induced by f on the f-invariant subspace V_i, we thus have that $p_i(f_i)^{e_i-1} = 0$. But this contradicts the fact that, by Theorem 3.2, the minimum polynomial of f_i is $p_i^{e_i}$. $\square$

Concerning the minimum polynomial of a skew-adjoint mapping, we have the following description.

Theorem 10.5

Let V be a non-zero finite-dimensional real inner product space and $f : V \to V$ a skew-adjoint linear mapping. If p is an irreducible factor of the minimum polynomial of f then either $p = X$ or $p = X^2 + b^2$ for some $b \neq 0$.

Proof

Since skew-adjoint mappings are clearly normal it follows by Theorem 10.4 that m_f is of the form $\prod_{i=1}^{k} p_i$ where $p_1, \ldots, p_k$ are distinct monic irreducible polynomials. We also know, by Theorem 10.1, that each p_i is either linear or of the form $(X - a_i)^2 + b_i^2$ with $b_i \neq 0$.

Suppose first that p_i is linear, say $p_i = X - a_i$, and let f_i be the mapping induced by f on the primary component $V_i = \operatorname{Ker} p_i(f)$. Then we have $f_i = a_i \operatorname{id}_{V_i}$ and consequently $f_i^\star = f_i$. Since f_i is also skew-adjoint, it follows that $0 = f_i = a_i \operatorname{id}_{V_i}$. Hence $a_i = 0$ and we have $p_i = X$.

Suppose now that p_i is not linear. Then we have

$$0 = p_i(f_i) = f_i^2 - 2a_i f_i + (a_i^2 + b_i^2)\operatorname{id}_{V_i}.$$

Since f_i is skew-adjoint, we deduce that

$$0 = f_i^2 + 2a_i f_i + (a_i^2 + b_i^2)\operatorname{id}_{V_i}.$$

These equalities give $4a_i f_i = 0$. Now $f_i \neq 0$ since otherwise we would have $p_i = m_{f_i} = X$, in contradiction to the hypothesis. Hence we must have $a_i = 0$ whence $p_i = X^2 + b_i^2$ where $b_i \neq 0$. $\square$

Corollary

If f is skew-adjoint then its minimum polynomial m_f is given as follows:

    (1) if $f = 0$ then

$$m_f = X;$$

    (2) if f is invertible then

$$m_f = \prod_{i=1}^{k}(X^2 + b_i^2)$$

for distinct non-zero real numbers $b_1, \ldots, b_k$;

    (3) if f is neither 0 nor invertible then

$$m_f = X \prod_{i=2}^{k}(X^2 + b_i^2)$$

for distinct non-zero real numbers $b_2, \ldots, b_k$.

Proof

This is immediate from Theorems 10.4 and 10.5 on recalling that f is invertible if and only if the constant term in m_f is non-zero. $\square$

We now observe how orthogonality enters the picture.

Theorem 10.6

If V is a non-zero finite-dimensional real inner product space and if $f : V \to V$ is skew-adjoint then the primary components of f are pairwise orthogonal.

Proof

Let V_i and V_j be primary components of f with $i \neq j$. With the usual notation, if f_i, f_j are the mappings induced on V_i, V_j by f suppose first that

$$m_{f_i} = X^2 + b_i^2, \quad m_{f_j} = X^2 + b_j^2$$

where b_i, b_j are non-zero. Observe first that, by Theorem 10.4, we have $b_i^2 \neq b_j^2$. Then, for $x_i \in V_i$ and $x_j \in V_j$,

$$
\begin{aligned}
0 &= \langle (f_i^2 + b_i^2 \mathrm{id}_{V_i})(x_i) \mid x_j \rangle \\
&= \langle f^2(x_i) \mid x_j \rangle + b_i^2 \langle x_i \mid x_j \rangle \\
&= \langle f(x_i) \mid -f(x_j) \rangle + b_i^2 \langle x_i \mid x_j \rangle \\
&= \langle x_i \mid f^2(x_j) \rangle + b_i^2 \langle x_i \mid x_j \rangle \\
&= \langle x_i \mid -b_j^2 x_j \rangle + b_i^2 \langle x_i \mid x_j \rangle \\
&= (b_i^2 - b_j^2) \langle x_i \mid x_j \rangle.
\end{aligned}
$$

Since $b_i^2 \neq b_j^2$, it follows that $\langle x_i \mid x_j \rangle = 0$.

Suppose now that $m_{f_i} = X^2 + b_i^2$ with $b_i \neq 0$, and $m_{f_j} = X$. In this case the above array becomes

$$
\begin{aligned}
0 &= \langle (f_i^2 + b_i^2 \mathrm{id}_{V_i})(x_i) \mid x_j \rangle \\
&= \langle f^2(x_i) \mid x_j \rangle + b_i^2 \langle x_i \mid x_j \rangle \\
&= \langle f(x_i) \mid -f(x_j) \rangle + b_i^2 \langle x_i \mid x_j \rangle \\
&= \langle f_i(x_i) \mid -f_j(x_j) \rangle + b_i^2 \langle x_i \mid x_j \rangle \\
&= \langle f_i(x_i) \mid 0_V \rangle + b_i^2 \langle x_i \mid x_j \rangle \\
&= b_i^2 \langle x_i \mid x_j \rangle
\end{aligned}
$$

whence again we see that $\langle x_i \mid x_j \rangle = 0$. $\square$

In order to establish the main theorem concerning skew-adjoint mappings, we require one further result.

Theorem 10.7

Let V be a non-zero finite-dimensional inner product space and let W be a subspace of V. Then W is f-invariant if and only if $W^\perp$ is f^*-invariant.

Proof

Suppose that W is f-invariant. Since $V = W \oplus W^\perp$ we have

$$(\forall x \in W)(\forall y \in W^\perp) \qquad \langle x \,|\, f^*(y)\rangle = \langle f(x)\,|\, y\rangle = 0.$$

It follows that $f^*(y) \in W^\perp$ for all $y \in W^\perp$ and so $W^\perp$ is f^*-invariant. Applying this observation again, we obtain the converse; for if $W^\perp$ is f^*-invariant then $W = W^{\perp\perp}$ is $f^{**} = f$-invariant. $\square$

Theorem 10.8

Let V be a non-zero finite-dimensional real inner product space and let $f : V \to V$ be a skew-adjoint linear mapping with minimum polynomial $m_f = X^2 + b^2$ where $b \neq 0$. Then $\dim V$ is even, and there is an ordered orthonormal basis of V with respect to which the matrix of f is of the form

$$
M(b) = \begin{bmatrix}
0 & -b & & & & & \\
b & 0 & & & & & \\
& & 0 & -b & & & \\
& & b & 0 & & & \\
& & & & \ddots & & \\
& & & & & 0 & -b \\
& & & & & b & 0
\end{bmatrix}.
$$

Proof

We begin by showing that $\dim V$ is even and that V is an orthogonal direct sum of f-cyclic subspaces each of dimension 2.

For this purpose, let y be a non-zero element of V. Observe that $f(y) \neq \lambda y$ for any scalar λ; for otherwise, since $f^2(y) = -b^2 y$, we would have $\lambda^2 = -b^2$ and hence the contradiction $b = 0$.

Let W_1 be the smallest f-invariant subspace containing y. Since $f^2(y) = -b^2 y$ it follows that W_1 is f-cyclic of dimension 2, a cyclic basis for W_1 being $\{y, f(y)\}$.

Consider now the decomposition $V = W_1 \oplus W_1^\perp$. This direct sum is orthogonal; for if p is the projection on W_1 parallel to $W_1^\perp$ then $\text{Im}\, p = W_1$ and $\text{Ker}\, p = W_1^\perp$ and so p is an ortho-projection. By Theorem 10.7, $W_1^\perp$ is f^*-invariant and so, since $f^* = -f$, we see that $W_1^\perp$ is also f-invariant, of dimension $\dim V - 2$.

Now let $V_1 = W_1^\perp$ and repeat the argument to obtain an orthogonal direct sum $V_1 = W_2 \oplus W_2^\perp$ of f-invariant subspaces in which W_2 is an f-cyclic subspace of dimension 2.

Continuing in this manner, we note that it is not possible in the final deomposition to have $\dim W_n^\perp = 1$. For, if this were so, then $W_n^\perp$ would have a singleton basis $\{z\}$ whence $f(z) \notin W_n^\perp$, a contradiction. Thus $W_n^\perp$ also has a basis of the form $\{z, f(z)\}$ and therefore is likewise f-cyclic of dimension 2. It follows therefore that $\dim V$ is even.

We now construct an orthonormal basis for each of the f-cyclic subspaces W_i. Consider the basis $\{y_i, f(y_i)\}$. Since

$$\|f(y_i)\|^2 = \langle f(y_i) \mid -f^*(y_i)\rangle = -\langle f^2(y_i) \mid y_i\rangle = b^2 \|y_i\|^2$$

it follows by applying the Gram–Schmidt process that an orthonormal basis for W_i is

$$B_i = \left\{ \frac{y_i}{\|y_i\|}, \frac{f(y_i)}{b\,\|y_i\|} \right\}.$$

Since $f^2(y_i) = -b^2 y_i$ it is readily seen that the matrix of f_i relative to B_i is

$$\begin{bmatrix} 0 & -b \\ b & 0 \end{bmatrix}.$$

Pasting together such bases, we obtain an orthonormal basis of V with respect to which the matrix of f is of the stated form. $\square$

Corollary 1

If V is a non-zero finite-dimensional real inner product space and $f : V \to V$ is a skew-adjoint linear mapping then there is an ordered orthonormal basis of V with respect to which the matrix of f is of the form

$$\begin{bmatrix} M_1 & & \\ & \ddots & \\ & & M_k \end{bmatrix}$$

in which each M_i is either 0 or as described in Theorem 10.8.

Proof

Combine the corollary of Theorem 10.5 with Theorems 10.6 and 10.8. $\square$

Corollary 2

A real square matrix is skew-symmetric if and only if it is orthogonally similar to a matrix of the form given in Corollary 1. $\square$

Example 10.2

The real skew-symmetric matrix

$$A = \begin{bmatrix} 0 & 1 & 1 \\ -1 & 0 & 1 \\ -1 & -1 & 0 \end{bmatrix}$$

has minimum polynomial $X(X^2 + 3)$ and so is orthogonally similar to the matrix

$$\begin{bmatrix} 0 & -\sqrt{3} & \\ \sqrt{3} & 0 & \\ & & 0 \end{bmatrix}.$$

EXERCISES

10.3 Show that the skew-symmetric matrix

$$A = \begin{bmatrix} 0 & 2 & -2 \\ -2 & 0 & -1 \\ 2 & 1 & 0 \end{bmatrix}$$

is orthogonally similar to the matrix

$$\begin{bmatrix} 0 & -3 & 0 \\ 3 & 0 & 0 \\ 0 & 0 & 0 \end{bmatrix}.$$

10.4 If A is a real skew-symmetric matrix prove that A and A' are orthogonally similar.

We now turn to the general problem of a normal linear mapping on a real inner product space.

For this purpose, recall from Theorem 10.2 that such a mapping can be expressed uniquely in the form $g + h$ where g is self-adjoint and h is skew-adjoint. Moreover, by Theorem 8.10, g is ortho-diagonalisable.

Theorem 10.9

Let V be a non-zero finite-dimensional real inner product space and let $f : V \to V$ be a normal linear mapping whose minimum polynomial is $m_f = (X-a)^2 + b^2$ where $b \neq 0$. If g, h are respectively the self-adjoint and skew-adjoint parts of f then

(1) *h is invertible;*
(2) *$m_g = X - a$;*
(3) *$m_h = X^2 + b^2$.*

Proof

(1) Suppose, by way of obtaining a contradiction, that $\operatorname{Ker} h \neq \{0_V\}$. Since f is normal we have $g \circ h = h \circ g$ by Theorem 10.2. Consequently, $\operatorname{Ker} h$ is g-invariant. Since $f = g + h$, the restriction of f to $\operatorname{Ker} h$ coincides with that of g. As $\operatorname{Ker} h$ is g-invariant, we can therefore define a linear mapping $f^+ : \operatorname{Ker} h \to \operatorname{Ker} h$ by the prescription

$$f^+(x) = f(x) = g(x).$$

Since g is self-adjoint, we see immediately that so is f^+. By Theorem 8.10, f^+ is then ortho-diagonalisable, and so its minimum polynomial is a product of distinct linear factors. But m_{f^+} must divide m_f which, by hypothesis, is of degree 2 and irreducible. This contradiction therefore gives $\operatorname{Ker} h = \{0_V\}$ whence h is invertible.

(2) Since $f = g + h$ with $g^* = g$ and $h^* = -h$ we have $f^* = g - h$. Since also, by hypothesis, $f^2 - 2af + (a^2 + b^2)\mathrm{id}_V = 0$ we have $(f^*)^2 - 2af^* + (a^2 + b^2)\mathrm{id}_V = 0$ and consequently

$$f^2 - (f^*)^2 = 2a(f - f^*) = 4ah.$$

Thus, since $f \circ f^* = f^* \circ f$, we see that

$$g \circ h = \tfrac{1}{2}(f + f^*) \circ \tfrac{1}{2}(f - f^*) = \tfrac{1}{4}[f^2 - (f^*)^2] = ah$$

and so $(g - a\mathrm{id}_V) \circ h = 0$. Since h is invertible by (1), we then have that $g - a\mathrm{id}_V = 0$ whence $m_g = X - a$.

(3) Since $f - h = g = a\mathrm{id}_V$ we have $f = h + a\mathrm{id}_V$ and so

$$\begin{aligned} 0 &= f^2 - 2af + (a^2 + b^2)\mathrm{id}_V \\ &= (h + a\mathrm{id}_V)^2 - 2a(h + a\mathrm{id}_V) + (a^2 + b^2)\mathrm{id}_V \\ &= h^2 + b^2\mathrm{id}_V. \end{aligned}$$

Since h is skew-adjoint and invertible it now follows by the Corollary to Theorem 10.5 that $m_h = X^2 + b^2$. $\square$

We can now extend Theorem 10.6 to normal mappings.

Theorem 10.10

If V is a non-zero finite-dimensional real inner product space and if $f : V \to V$ is normal then the primary components of f are pairwise orthogonal.

Proof

By Theorems 10.1 and 10.4 the minimum polynomial of f has the general form

$$m_f = (X - a_1) \prod_{i=2}^{k} (X^2 - 2a_i X + a_i^2 + b_i^2)$$

where each $b_i \neq 0$. The primary components of f are therefore $V_1 = \mathrm{Ker}\,(f - a_1\mathrm{id}_V)$ and

$$(i = 2, \ldots, k) \qquad V_i = \mathrm{Ker}\,\left(f^2 - 2a_i f + (a_i^2 + b_i^2)\mathrm{id}_V\right).$$

Moreover, for each i, the induced mapping f_i on V_i is normal with minimum polynomial $X - a_1$ if $i = 1$ and $X^2 - 2a_i X + a_i^2 + b_i^2$ otherwise.

Now $f_i = g_i + h_i$ where g_i is self-adjoint and h_i is skew-adjoint. Moreover, g_i and h_i coincide with the mappings induced on V_i by g and h. To see this, let these mappings be g', h' respectively. Then for every $x \in V_i$ we have

$$g_i(x) + h_i(x) = f_i(x) = f(x) = g(x) + h(x) = g'(x) + h'(x)$$

and so $g_i - g' = h' - h_i$. Since the left hand side is self-adjoint and the right hand side is skew-adjoint, each must be zero and we see that $g_i = g'$ and $h_i = h'$.

Suppose now that $i, j > 1$ with $i \neq j$. Then the minimum polynomials of f_i, f_j are

$$m_{f_i} = X^2 - 2a_iX + a_i^2 + b_i^2, \quad m_{f_j} = X^2 - 2a_jX + a_j^2 + b_j^2$$

where either $a_i \neq a_j$ or $b_i^2 \neq b_j^2$. Then, by Theorem 10.9, we have

$$m_{g_i} = X - a_i, \quad m_{g_j} = X - a_j, \quad m_{h_i} = X^2 + b_i^2, \quad m_{h_j} = X^2 + b_j^2.$$

Now given $x_i \in V_i$ and $x_j \in V_j$ we therefore have, precisely as in the proof of Theorem 10.6,

$$0 = \langle (h_i^2 + b_i^2 \mathrm{id}_{V_i})(x_i) \mid x_j \rangle = (b_i^2 - b_j^2)\langle x_i \mid x_j \rangle,$$

so that in the case where $b_i^2 \neq b_j^2$ we have $\langle x_i \mid x_j \rangle = 0$.

Likewise,

$$\begin{aligned}
0 = \langle (g_i - a_i \mathrm{id}_{V_i})(x_i) \mid x_j \rangle &= \langle g(x_i) \mid x_j \rangle - a_i \langle x_i \mid x_j \rangle \\
&= \langle x_i \mid g(x_j) \rangle - a_i \langle x_i \mid x_j \rangle \\
&= \langle x_i \mid g_j(x_j) \rangle - a_i \langle x_i \mid x_j \rangle \\
&= a_j \langle x_i \mid x_j \rangle - a_i \langle x_i \mid x_j \rangle \\
&= (a_j - a_i)\langle x_i \mid x_j \rangle,
\end{aligned}$$

so that in the case where $a_i \neq a_j$ we also have $\langle x_i \mid x_j \rangle = 0$.

We thus see that $V_2, \ldots, V_k$ are pairwise orthogonal. That V_1 is orthogonal to each V_i for $i \geq 2$ follows from the above equalities on taking $j = 1$ and using the fact that $f_1 = a_1 \mathrm{id}_{V_1}$ is self-adjoint and therefore $g_1 = f_1$ and $h_1 = 0$.   □

We can now establish the main result.

Theorem 10.11

If V is a non-zero finite-dimensional real inner product space and if $f : V \to V$ is a normal linear mapping then there is an ordered orthonormal basis of V with respect to which the matrix of f is of the form

$$\begin{bmatrix} A_1 & & & \\ & A_2 & & \\ & & \ddots & \\ & & & A_k \end{bmatrix}$$

where each A_i is either a 1×1 matrix or a 2×2 matrix of the form

$$\begin{bmatrix} \alpha & -\beta \\ \beta & \alpha \end{bmatrix}$$

in which $\beta \neq 0$.

Proof

With the same notation as above, let

$$m_f = (X - a_1) \prod_{i=2}^{k} (X^2 - 2a_iX + a_i^2 + b_i^2)$$

and let the primary components of f be V_i for $i = 1, \ldots, k$. Then $m_{f_i} = X - a_1$ if $i = 1$ and $m_{f_i} = X^2 - 2a_iX + a_i^2 + b_i^2$ otherwise.

For each V_i with $i \neq 1$ we have $f_i = g_i + h_i$ where the self-adjoint part g_i has minimum polynomial $X - a_i$ and the skew-adjoint part has minimum polynomial $X^2 + b_i^2$. Now by Theorem 10.8 there is an ordered orthonormal basis B_i of V_i with respect to which the matrix of h_i is

$$M(b_i) = \begin{bmatrix} 0 & -b_i & & & & & \\ b_i & 0 & & & & & \\ & & 0 & -b_i & & & \\ & & b_i & 0 & & & \\ & & & & \ddots & & \\ & & & & & 0 & -b_i \\ & & & & & b_i & 0 \end{bmatrix}.$$

Since the minimum polynomial of g_i is $X - a_i$ we have $g_i(x) = a_ix$ for every $x \in B_i$ and so the matrix of g_i with respect to B_i is a diagonal matrix all of whose entries are a_i. It now follows that the matrix of $f_i = g_i + h_i$ with respect to B_i is

$$M(a_i, b_i) = \begin{bmatrix} a_i & -b_i & & & & & \\ b_i & a_i & & & & & \\ & & a_i & -b_i & & & \\ & & b_i & a_i & & & \\ & & & & \ddots & & \\ & & & & & a_i & -b_i \\ & & & & & b_i & a_i \end{bmatrix}.$$

In the case where $i = 1$, we have $f_1 = a_1 \mathrm{id}_{V_1}$ so f_1 is self-adjoint. By Theorem 8.10, there is an ordered orthonormal basis with respect to which the matrix of f_1 is diagonal.

Now since, by Theorem 10.10, the primary components are pairwise orthogonal we can paste together the ordered orthonormal bases in question, thereby obtaining an ordered orthonormal basis of V with respect to which the matrix of f is of the form stated. □

Corollary

A real square matrix is normal if and only if it is orthogonally similar to a matrix of the form described in Theorem 10.11.   □

Our labours produce a bonus. Recall that if V is a real inner product space then a linear mapping $f : V \rightarrow V$ is **orthogonal** if and only if f^{-1} exists and is $f^{\star}$. Likewise, a real square matrix A is **orthogonal** if A^{-1} exists and is A'. Clearly, an orthogonal mapping commutes with its adjoint and is therefore normal. Correspondingly, an orthogonal matrix commutes with its transpose. We can therfore deduce as a special case of Theorem 10.11 a canonical form for orthogonal mappings and matrices.

Theorem 10.12

If V is a non-zero finite-dimensional real inner product space and if $f : V \rightarrow V$ is an orthogonal linear mapping then there is an ordered orthonormal basis of V with respect to which the matrix of f is of the form

$$\begin{bmatrix} I_m & & & & & \\ & -I_p & & & & \\ & & A_2 & & & \\ & & & A_3 & & \\ & & & & \ddots & \\ & & & & & A_k \end{bmatrix}$$

in which each A_i is a 2×2 matrix of the form

$$\begin{bmatrix} \alpha & -\beta \\ \beta & \alpha \end{bmatrix}$$

where $\beta \neq 0$ and $\alpha^2 + \beta^2 = 1$.

Proof

With the same notation as in the above, we have that the matrix $M(a_i, b_i)$ which represents f_i relative to the ordered basis B_i is an orthogonal matrix (since f_i is orthogonal). Multiplying this matrix by its transpose we obtain an identity matrix and, equating entries, we see that $a_i^2 + b_i^2 = 1$. As for the primary component V_1, the matrix of f_1 is diagonal. Since the square of this diagonal matrix is an identity matrix, its entries must be ± 1. We can now rearrange the basis in such a way that the matrix of f has the stated form. $\square$

Example 10.3

If $f : \mathbb{R}^3 \rightarrow \mathbb{R}^3$ is orthogonal then f is called a **rotation** if $\det A = 1$ for any matrix A that represents f. If f is a rotation then there is an ordered orthonormal basis of $\mathbb{R}^3$ with respect to which the matrix of f is

$$\begin{bmatrix} 1 & 0 & 0 \\ 0 & \cos \vartheta & -\sin \vartheta \\ 0 & \sin \vartheta & \cos \vartheta \end{bmatrix}$$

for some real number ϑ.

11
Computer Assistance

Many applications of linear algebra require careful, and sometimes rather tedious, calculations by hand. As the reader will be aware, these can often be subject to error. The use of a computer is therefore called for. As far as computation in algebra is concerned, there are several packages that have been developed specifically for this purpose. In this chapter we give a brief introduction, by way of a tutorial, to the package 'LinearAlgebra' in MAPLE 7. Having mastered the techniques, the reader may freely check some of the answers to previous questions!

Having opened MAPLE, begin with the input:

```
> with(LinearAlgebra):
```

(1) *Matrices*

There are several different ways to input a matrix. Here is the first, which merely gives the matrix as a list of its rows (the matrix palette may also be used to do this). At each stage the MAPLE output is generated immediately following the semi-colon on pressing the ENTER key.

input :

```
> m1:=Matrix([[1,2,3],[2,3,-1],[6,-3,-4]]);
```

output :

$$m1 := \begin{bmatrix} 1 & 2 & 3 \\ 2 & 3 & -1 \\ 6 & -3 & -4 \end{bmatrix}$$

In order to illustrate how to do matrix algebra with MAPLE, let us input another matrix of the same size:

```
> m2:=Matrix([[-1,4,7],[-2,5,41],[-6,-3,3]]);
```

$$m2 := \begin{bmatrix} -1 & 4 & 7 \\ -2 & 5 & 41 \\ -6 & -3 & 3 \end{bmatrix}$$

Then here is one way of adding matrices, using the 'Add' command:

```
> m3:=Add(m1,m2);
```

$$m3 := \begin{bmatrix} 0 & 6 & 10 \\ 0 & 8 & 40 \\ 0 & -6 & -1 \end{bmatrix}$$

As for multiplying matrices, this can be achieved by using the 'Multiply' command. To multiply the above matrices, input:

```
> m4:=Multiply(m1,m2);
```

$$m4 := \begin{bmatrix} -23 & 5 & 98 \\ -2 & 26 & 134 \\ 24 & 21 & -93 \end{bmatrix}$$

Now 'Add' also allows linear combinations to be computed. Here, for example, is how to obtain $3m1 + 4m2$:

```
> Add(m1,m2,3,4);
```

$$\begin{bmatrix} -1 & 22 & 37 \\ -2 & 29 & 161 \\ -6 & -21 & 0 \end{bmatrix}$$

(2) A simpler method

An more convenient way to input commands is to use algebraic operations. By way of example:

```
> m1+m2;
```

$$\begin{bmatrix} 0 & 6 & 10 \\ 0 & 8 & 40 \\ 0 & -6 & -1 \end{bmatrix}$$

Multiplication by scalars is obtained by using a '$\star$':

```
> 3*m1+4*m2;
```

$$\begin{bmatrix} -1 & 22 & 37 \\ -2 & 29 & 161 \\ -6 & -21 & 0 \end{bmatrix}$$

Multiplication of matrices is obtained by using a '.' :

```
> m1.m2;
```

$$\begin{bmatrix} -23 & 5 & 98 \\ -2 & 26 & 134 \\ 24 & 21 & -93 \end{bmatrix}$$

As for the more complicated expression: $m1(4m2 - 5m1^2)$:

```
> m1.(4*m2-5*m1^2);
```

$$\begin{bmatrix} 133 & -360 & -178 \\ -388 & -51 & 726 \\ -1044 & 654 & 803 \end{bmatrix}$$

(3) Inverses
Inverses of matrices can be achieved by using either 'MatrixInverse' or as follows (here it is necessary to insert brackets round the -1):

```
> m1^(-1);
```

$$
\begin{bmatrix}
\frac{15}{83} & \frac{1}{83} & \frac{11}{83} \\[4pt]
\frac{-2}{83} & \frac{22}{83} & \frac{-7}{83} \\[4pt]
\frac{24}{83} & \frac{-15}{83} & \frac{1}{83}
\end{bmatrix}
$$

(4) *Determinants*

To compute a determinant, use the command 'Determinant':

```
> Determinant(m1);
```

$$-83$$

Of course the determinant of a product is the product of the determinants: each of the commands

```
> Determinant(m1.m2);
> Determinant(m1)*Determinant(m2);
```

gives

$$70218$$

Note that we can use negative powers in products:

```
> m1^(-3).m2^3;
```

$$
\begin{bmatrix}
\frac{1841635}{571787} & \frac{4640885}{571787} & \frac{9128894}{571787} \\[6pt]
\frac{-19800794}{571787} & \frac{-32833132}{571787} & \frac{-62530138}{571787} \\[6pt]
\frac{8866674}{571787} & \frac{19933848}{571787} & \frac{45220050}{571787}
\end{bmatrix}
$$

(5) *More on defining matrices*

We now look at other ways of defining a matrix. We start with a clean sheet (to remove all previous definitions):

```
> restart;
with(LinearAlgebra):
```

We can enter a matrix as a row of columns:

```
M0:= <<a,b,c>|<d,e,f>|<g,h,i>>;
```

$$
M0 := \begin{bmatrix}
a & d & g \\
b & e & h \\
c & f & i
\end{bmatrix}
$$

or as a column of rows (this can also be done using the matrix palette):

```
M1:= <<a | b | c> , <d | e | f> , <g | h | i>>;
```

$$M1 := \begin{bmatrix} a & b & c \\ d & e & f \\ g & h & i \end{bmatrix}$$

Then, for example, we have

```
> M1^2;
```

$$\begin{bmatrix} a^2 + bd + cg & ab + be + ch & ac + bf + ci \\ da + ed + fg & bd + e^2 + fh & dc + ef + fi \\ ga + hd + ig & gb + hc + ih & cg + fh + i^2 \end{bmatrix}$$

```
> Determinant(M1);
```

$$aei - afh + dch - dbi + gbf - gec$$

Particular types of matrix can be dealt with as follows. A 3×3 lower triangular matrix, for example:

```
> M2:=Matrix(3,[[1],[2,3],[4,5,6]],
shape=triangular[lower]);
```

$$M2 := \begin{bmatrix} 1 & 0 & 0 \\ 2 & 3 & 0 \\ 4 & 5 & 6 \end{bmatrix}$$

For a symmetric 3×3 matrix, begin with

```
> M3:=Matrix(3,3,shape=symmetric);
```

$$M3 := \begin{bmatrix} 0 & 0 & 0 \\ 0 & 0 & 0 \\ 0 & 0 & 0 \end{bmatrix}$$

then input, for example,

```
> M3[1,1]:=2;M3[1,3]:=23;M3[2,3]:=Pi;
```

$$M3_{1,1} := 2$$

$$M3_{1,3} := 23$$

$$M3_{2,3} := \pi$$

```
> M3;
```

$$M3 := \begin{bmatrix} 2 & 0 & 23 \\ 0 & 0 & \pi \\ 23 & \pi & 0 \end{bmatrix}$$

```
> Determinant (M3);
```

$$-2\pi^2$$

Skew-symmetric matrices can be done similarly:

```
> M4:=Matrix(3,3,shape=antisymmetric);
```

$$M4 := \begin{bmatrix} 0 & 0 & 0 \\ 0 & 0 & 0 \\ 0 & 0 & 0 \end{bmatrix}$$

```
> M4[1,2]:=2; M4[1,3]:=23;M4[2,3]:=Pi;
```

$$M4_{1,2} := 2$$

$$M4_{1,3} := 23$$

$$M4_{2,3} := \pi$$

```
> M4;
```

$$M4 := \begin{bmatrix} 0 & 2 & 23 \\ -2 & 0 & \pi \\ -23 & -\pi & 0 \end{bmatrix}$$

```
> Determinant (M4);
```

$$0$$

Hermitian matrices can be dealt with as follows (note here the $\star$):

```
> M5:=Matrix(3,3,shape=hermitian);
```

$$M5 := \begin{bmatrix} 0 & 0 & 0 \\ 0 & 0 & 0 \\ 0 & 0 & 0 \end{bmatrix}$$

```
> M5[1,1]:=2; M5[1,2]:=5+7*I; M5[1,3]:=23-6*I;
M5[2,3]:=I;
```

$$M5_{1,1} := 2$$

$$M5_{1,2} := 5 + 7I$$

$$M5_{1,3} := 23 - 6I$$

$$M5_{2,3} := I$$

```
> M5;
```

$$M5 := \begin{bmatrix} 5 + 7I & 0 & 23 - 6I \\ 5 - 7I & 0 & I \\ 23 + 6I & -I & 0 \end{bmatrix}$$

```
> Determinant(M5);
```

$$-384$$

Submatrices can be defined, for example, as follows:

```
> m1:=M5[2..3,2..3];
```

$$m1 := \begin{bmatrix} 0 & I \\ -I & 0 \end{bmatrix}$$

We can also input matrices where the (i,j)-th entry is a function of i and j:

```
> m2:=Matrix(6,6,(i,j)->i*j);
```

$$m2 := \begin{bmatrix} 1 & 2 & 3 & 4 & 5 & 6 \\ 2 & 4 & 6 & 8 & 10 & 12 \\ 3 & 6 & 9 & 12 & 15 & 18 \\ 4 & 8 & 12 & 16 & 20 & 24 \\ 5 & 10 & 15 & 20 & 25 & 30 \\ 6 & 12 & 18 & 24 & 30 & 36 \end{bmatrix}$$

(6) *Writing procedures*

To use a more complicated function we write a procedure. The following example illustrates a very simple procedure in order to define the 6×6 identity matrix:

```
> f:=proc(i,j);
if i=j then 1 else 0 fi;
end;
```

$$f := \mathbf{proc}\,(i, j)\,\mathbf{if}\ i = j\ \mathbf{then}\ 1\ \mathbf{else}\ 0\ \mathbf{fi}\ \mathbf{end}$$

```
> m3:=Matrix(6,6,f);
```

$$m3 := \begin{bmatrix} 1 & 0 & 0 & 0 & 0 & 0 \\ 0 & 1 & 0 & 0 & 0 & 0 \\ 0 & 0 & 1 & 0 & 0 & 0 \\ 0 & 0 & 0 & 1 & 0 & 0 \\ 0 & 0 & 0 & 0 & 1 & 0 \\ 0 & 0 & 0 & 0 & 0 & 1 \end{bmatrix}$$

Here is a more complicated procedure:

```
> f:=proc(i,j);
if i> j then x else if i=j then 0 else -y fi fi;
end;
```

$$f := \mathbf{proc}\,(i,\,j)\,\mathbf{if}\ j < i\ \mathbf{then}\ x\ \mathbf{else}\ \mathbf{if}\ i = j\ \mathbf{then}\ 0\ \mathbf{else}\ -y\ \mathbf{fi}\ \mathbf{fi}\ \mathbf{end}$$

The 6 x 6 matrix whose (i,j)-th entry is given by this function is:

```
> m4:=Matrix(6,6,f);
```

$$m4 := \begin{bmatrix} 0 & -y & -y & -y & -y & -y \\ x & 0 & -y & -y & -y & -y \\ x & x & 0 & -y & -y & -y \\ x & x & x & 0 & -y & -y \\ x & x & x & x & 0 & -y \\ x & x & x & x & x & 0 \end{bmatrix}$$

```
> d:=Determinant(m4);
```

$$d := (x^4 - x^3y + y^2x^2 - y^3x + y^4)xy$$

We can simplify this:

```
> expand(d*(x+y));
```

$$x^6y + xy^6$$

So the determinant of m4 is $\dfrac{x^6y + xy^6}{x + y}$. Examine this for a few values of n:

```
> for n from 2 to 8 do
m4:=Matrix(n,n,f);
d:=Determinant(m4);
print(' size', n, ' gives',expand(d*(x+y)));
od:
```

$$size, 2, \; gives, yx^2 + y^2x$$
$$size, 3, \; gives, -x^3y + y^3x$$
$$size, 4, \; gives, x^4y + y^4x$$
$$size, 5, \; gives, -x^5y + y^5x$$
$$size, 6, \; gives, x^6y + xy^6$$
$$size, 7, \; gives, -x^7y + xy^7$$
$$size, 8, \; gives, x^8y + xy^8$$

The general theorem should be easy to spot. Try to prove it.

Here is a way to input a tri-diagonal matrix using the command 'BandMatrix':

```
> m5:=BandMatrix([sin(x),x^3,-cos(x)],1,6,6);
```

$$m5 := \begin{bmatrix} x^3 & -\cos(x) & 0 & 0 & 0 & 0 \\ \sin(x) & x^3 & -\cos(x) & 0 & 0 & 0 \\ 0 & \sin(x) & x^3 & -\cos(x) & 0 & 0 \\ 0 & 0 & \sin(x) & x^3 & -\cos(x) & 0 \\ 0 & 0 & 0 & \sin(x) & x^3 & -\cos(x) \\ 0 & 0 & 0 & 0 & \sin(x) & x^3 \end{bmatrix}$$

```
> d:=Determinant(m5);
```

$$d := x^{18} + 5x^{12}\sin(x)\cos(x) + 6x^6\cos(x)^2\sin(x)^2 + \cos(x)^3\sin(x)^3$$

We can also differentiate the entries of a matrix using the 'map' command:

```
> m6:=map(diff,m5,x);
```

$$m6 := \begin{bmatrix} 3x^2 & \sin(x) & 0 & 0 & 0 & 0 \\ \cos(x) & 3x^2 & \sin(x) & 0 & 0 & 0 \\ 0 & \cos(x) & 3x^2 & \sin(x) & 0 & 0 \\ 0 & 0 & \cos(x) & 3x^2 & \sin(x) & 0 \\ 0 & 0 & 0 & \cos(x) & 3x^2 & \sin(x) \\ 0 & 0 & 0 & 0 & \cos(x) & 3x^2 \end{bmatrix}$$

```
> d1:=Determinant(m6);
```

$$d1 := 729x^{12} - 405x^8\sin(x)\cos(x) + 54x^4\cos(x)^2\sin(x)^2 - \cos(x)^3\sin(x)^3$$

(7) *More on determinants*

Again start with a clean sheet:

```
> restart;
with(LinearAlgebra):
```

We now examine the determinants of some other matrices and try to spot the general results. Here is the first of two examples:

```
> f:=proc(i,j);
if i = j then x else b fi;
end;
```

$$f := \mathbf{proc}\,(i,j)\,\mathbf{if}\; i = j \;\mathbf{then}\; x \;\mathbf{else}\; b \;\mathbf{fi}\;\mathbf{end}$$

Consider a 6 x 6 matrix whose entries are given by this function:

```
> m:=Matrix(6,6,f);
```

$$m := \begin{bmatrix} x & b & b & b & b & b \\ b & x & b & b & b & b \\ b & b & x & b & b & b \\ b & b & b & x & b & b \\ b & b & b & b & x & b \\ b & b & b & b & b & x \end{bmatrix}$$

```
> d:=Determinant(m);
```

$$d := x^6 - 15x^4b^2 + 40x^3b^3 - 45b^4x^2 + 24b^5x - 5b^6$$

Factorise the answer:

```
> factor(d);
```

$$(x + 5b)(x - b)^5$$

This looks like a simple result. Let us examine a few cases:

```
> for n from 2 to 8 do
m:=Matrix(n,n,f);
d:=Determinant(m);
print(factor(d));
od:
```

$$(x - b)(x + b)$$
$$(x + 2b)(x - b)^2$$
$$(x + 3b)(x - b)^3$$
$$(x + 4b)(x - b)^4$$
$$(x + 5b)(x - b)^5$$
$$(x + 6b)(x - b)^6$$
$$(x + 7b)(x - b)^7$$

It is now easy to spot the general theorem. Try to prove it.

Here is a second example:

```
> f:=proc(i,j);
if i = j then x else if i < j then y else -y fi fi;
end;
```

$$f := \mathbf{proc}(i,j) \, \mathbf{if} \, i = j \, \mathbf{then} \, x \, \mathbf{else} \, \mathbf{if} \, i < j \, \mathbf{then} \, y \, \mathbf{else} \, -y \, \mathbf{fi} \, \mathbf{fi} \, \mathbf{end}$$

```
> m:=Matrix(6,6,f);
```

$$m := \begin{bmatrix} x & y & y & y & y & y \\ -y & x & y & y & y & y \\ -y & -y & x & y & y & y \\ -y & -y & -y & x & y & y \\ -y & -y & -y & -y & x & y \\ -y & -y & -y & -y & -y & x \end{bmatrix}$$

```
> d:=Determinant(m);
```

$$d := x^6 + 15x^4y^2 + 15y^4x^2 + y^6$$

It is less easy to spot the general result here. Let us try a few cases:

```
> for n from 2 to 8 do
m:=Matrix(n,n,f);
d:=Determinant(m);
print(sort(d,[x,y]));
od:
```

$$x^2 + y^2$$
$$x^3 + 3xy^2$$
$$x^4 + 6x^2y^2 + y^4$$
$$x^5 + 10x^3y^2 + 5xy^4$$
$$x^6 + 15x^4y^2 + 15x^2y^4 + y^6$$
$$x^7 + 21x^5y^2 + 35x^3y^4 + 7xy^6$$
$$x^8 + 28x^6y^2 + 70x^4y^4 + 28x^2y^6 + y^8$$

Here it is harder to spot a simple form of the solution. However there is a compact form for the solution:

```
> for i from 2 to 8 do
expand(((x+y)^i+(x-y)^i)/2);
od;
```

$$x^2 + y^2$$
$$x^3 + 3xy^2$$
$$x^4 + 6x^2y^2 + y^4$$
$$x^5 + 10x^3y^2 + 5xy^4$$
$$x^6 + 15x^4y^2 + 15x^2y^4 + y^6$$
$$x^7 + 21x^5y^2 + 35x^3y^4 + 7xy^6$$
$$x^8 + 28x^6y^2 + 70x^4y^4 + 28x^2y^6 + y^8$$

Now try to prove the general result.

(8) *Matrices with subscripted entries*

Dealing with these is easy. Consider the Vandermonde matrix

```
> m:=Matrix(4,4,(i,j)-> b[j]^(i-1));
```

$$m := \begin{bmatrix} 1 & 1 & 1 & 1 \\ b_1 & b_2 & b_3 & b_4 \\ b_1^2 & b_2^2 & b_3^2 & b_4^2 \\ b_1^3 & b_2^3 & b_3^3 & b_4^3 \end{bmatrix}$$

```
> d:=Determinant(m);
```

$$d := b_2b_3^2b_4^3 - b_2b_4^2b_3^3 - b_2^2b_3b_4^3 + b_2^2b_4b_3^3 + b_2^3b_3b_4^2 - b_2^3b_4b_3^2 - b_1b_3^2b_4^3$$
$$+ b_1b_4^2b_3^3 + b_1b_2^2b_4^3 - b_1b_2^2b_3^3 - b_1b_2^3b_4^2 + b_1b_2^3b_3^2 + b_1^2b_3b_4^3 - b_1^2b_4b_3^3$$
$$- b_1^2b_2b_4^3 + b_1^2b_2b_3^3 + b_1^2b_2^3b_4 - b_1^2b_2^3b_3 - b_1^3b_3b_4^2 + b_1^3b_4b_3^2 + b_1^3b_2b_4^2$$
$$- b_1^3b_2b_3^2 - b_1^3b_2^2b_4 + b_1^3b_2^2b_3$$

No surprise what we will obtain when we factorise this:

```
> factor(d);
```

$$(-b_4 + b_3)(b_2 - b_3)(b_2 - b_4)(-b_3 + b_1)(b_1 - b_4)(b_1 - b_2)$$

(9) *Eigenvalues and eigenvectors*

Let us now find the eigenvalues of a matrix. First input a matrix:

```
> a:=Matrix([[-2,-3,-3],[-1,0,-1],[5,5,6]]);
```

$$a := \begin{bmatrix} -2 & -3 & -3 \\ -1 & 0 & -1 \\ 5 & 5 & 6 \end{bmatrix}$$

We use the command 'Eigenvalues':

```
> e:=Eigenvalues(a);
```

$$e := \begin{bmatrix} 2 \\ 1 \\ 1 \end{bmatrix}$$

```
> e[1];
```

$$2$$

```
> e[2];
```

$$1$$

```
> e:=Eigenvalues(a, output='Vector[row]');
```

$$e := [2, 1, 1]$$

Of course we can find the eigenvalues as the roots of the characteristic polynomial:

```
> ch:=CharacteristicPolynomial(a,X);
```

$$ch := 5X - 4X^2 - 2 + X^3$$

```
> factor(ch);
```

$$(X - 2)(X - 1)^2$$

We can also find the eigenvalues as the solutions of the characteristic equation:

```
> solve(ch=0,X);
```

$$2, 1, 1$$

Next, we can find the eigenvectors with the 'Eigenvectors' command:

```
> ev1:=Eigenvectors(a);
```

$$ev1 := \begin{bmatrix} 1 \\ 1 \\ 2 \end{bmatrix}, \begin{bmatrix} 0 & 1 & 1 \\ 1 & 0 & \frac{1}{3} \\ -1 & -1 & -\frac{5}{3} \end{bmatrix}$$

```
> ev2:=Eigenvectors(a, output='list');
```

$$ev2 := \left[\left[1,2\left\{\begin{bmatrix}1\\0\\-1\end{bmatrix}, \begin{bmatrix}0\\1\\-1\end{bmatrix}\right\}\right], \left[2,1,\left\{\begin{bmatrix}3\\1\\5\end{bmatrix}\right\}\right]\right]$$

This needs some interpretation. It returns a list of lists. Each list has the form $[ei, mi, \{v[1, i], ...v[ni, i]\}]$ where ei is the eigenvalue, mi is its algebraic multiplicity and the set $\{v[1, i], ..., v[ni, i]\}$ gives ni linearly independent eigenvectors where ni is the geometric multiplicity. Hence the eigenvalue 1 has algebraic multiplicity 2 with

$$\begin{bmatrix}1\\0\\-1\end{bmatrix}, \begin{bmatrix}0\\1\\-1\end{bmatrix}$$

as two linearly independent eigenvectors.

The corresponding eigenspaces can also be found with a 'NullSpace' command:

```
> id:=IdentityMatrix(3):
> k1:=NullSpace(a-1*id);
```

$$k1 := \left\{\begin{bmatrix}-1\\0\\1\end{bmatrix}, \begin{bmatrix}-1\\1\\0\end{bmatrix}\right\}$$

```
> k2:=NullSpace(a-2*id);
```

$$k2 := \left\{\begin{bmatrix}3\\1\\5\end{bmatrix}\right\}$$

```
> p:=Matrix([k1[1],k1[2],k2[1]]);
```

$$p := \begin{bmatrix}-1 & -1 & 3\\0 & 1 & 1\\1 & 0 & -5\end{bmatrix}$$

Now compute $p^{-1}ap$ and we should obtain a diagonal matrix:

```
> p^(-1).a.p;
```

$$\begin{bmatrix}1 & 0 & 0\\0 & 1 & 0\\0 & 0 & 2\end{bmatrix}$$

(10) *Computing Jordan normal forms*

Exercise 5.4 asks for the Jordan normal form of the following matrix:

```
> A:=Matrix([[5,-1,-3,2,-5],[0,2,0,0,0],[1,0,1,1,-2],
[0,-1,0,3,1],[1,-1,-1,1,1]]);
```

$$A := \begin{bmatrix} 5 & -1 & -3 & 2 & -5 \\ 0 & 2 & 0 & 0 & 0 \\ 1 & 0 & 1 & 1 & -2 \\ 0 & -1 & 0 & 3 & 1 \\ 1 & -1 & -1 & 1 & 1 \end{bmatrix}$$

First compute the characteristic polynomial:

```
> factor(CharacteristicPolynomial(A,X));
```

$$(-3 + X)^2(X - 2)^3$$

and then the minimum polynomial:

```
> factor(MinimalPolynomial(A,X));
```

$$(X - 2)^2(-3 + X)^2$$

Now we could write down the Jordan normal form, but MAPLE does it easily:

```
> JordanForm(A);
```

$$\begin{bmatrix} 2 & 1 & 0 & 0 & 0 \\ 0 & 2 & 0 & 0 & 0 \\ 0 & 0 & 3 & 1 & 0 \\ 0 & 0 & 0 & 3 & 0 \\ 0 & 0 & 0 & 0 & 2 \end{bmatrix}$$

To determine the transition matrix P to a Jordan basis:

```
> P:=JordanForm(A,output='Q');;
```

$$\begin{bmatrix} 1 & 1 & -1 & 3 & 3 \\ 0 & 1 & 0 & 0 & 1 \\ 1 & 1 & 0 & 0 & 1 \\ 0 & 1 & 1 & -1 & 0 \\ 0 & 0 & 0 & 1 & 1 \end{bmatrix}$$

A knowledge of P allows the determination of a corresponding Jordan basis.

(11) *The Gram–Schmidt process*

Let us see how to apply the Gram–Schmidt orthonormalisation process. Consider again Example 1.12.

```
> x1:=<0|1|1>; x2:=<1|0|1>; x3:=<1|1|0>;
```

$$x1 := [0, 1, 1]$$
$$x2 := [1, 0, 1]$$
$$x3 := [1, 1, 0]$$

```
> GramSchmidt([x1,x2,x3]);
```

$$[[0, 1, 1], [1 - \tfrac{1}{2}, \tfrac{1}{2}], [\tfrac{2}{3}, \tfrac{2}{3}, -\tfrac{2}{3}]]$$

Now we normalize to find an orthonormal basis:

```
> GramSchmidt([x1,x2,x3],normalized);
```

$$[[0, \tfrac{1}{2}\sqrt{2}, \tfrac{1}{2}\sqrt{2}], [\tfrac{1}{3}\sqrt{6}, -\tfrac{1}{6}\sqrt{6}, \tfrac{1}{6}\sqrt{6}], [\tfrac{1}{3}\sqrt{3}, \tfrac{1}{3}\sqrt{3}, -\tfrac{1}{3}\sqrt{3}]]$$

We can compute inner products using the 'DotProduct' command:

```
> DotProduct(<1/2|37/11|-15/17|7/9>,
<1/11|3/5|2/7|9/16>);
```

$$\frac{235519}{104720}$$

and so we can apply the Gram–Schmidt process as in hand calculations:

```
> y1:=Normalize(x1,Euclidean);
```

$$y1 := [0, \tfrac{1}{2}\sqrt{2}, \tfrac{1}{2}\sqrt{2}]$$

```
> y2:=Normalize(x2-DotProduct(x2,y1)*y1,Euclidean);
```

$$y2 := [\tfrac{1}{3}\sqrt{6}, -\tfrac{1}{6}\sqrt{6}, \tfrac{1}{6}\sqrt{6}]$$

```
> y3:=Normalize(x3-DotProduct(x3,y2)*y2
-DotProduct(x3,y1)*y1,Euclidean);
```

$$y3 := [\tfrac{1}{3}\sqrt{3}, \tfrac{1}{3}\sqrt{3}, -\tfrac{1}{3}\sqrt{3}]$$

EXERCISES

11.1 Find the eigenvalues and corresponding eigenvectors of

$$A = \begin{bmatrix} 1 & 2 & 3 & 4 \\ 5 & 6 & 7 & 8 \\ 9 & 10 & 11 & 12 \\ 13 & 14 & 15 & 16 \end{bmatrix}.$$

11.2 Find the determinant and the inverse of

$$A = \begin{bmatrix} 1 & 2 & 3 & 4 & 5 & 6 \\ 9 & 8 & 7 & 7 & 8 & 9 \\ 1 & 3 & 5 & 7 & 2 & 4 \\ 5 & 4 & 5 & 6 & 5 & 6 \\ 2 & 9 & 2 & 7 & 2 & 9 \\ 3 & 6 & 4 & 5 & 6 & 7 \end{bmatrix}.$$

11.3 For which value of $x \in \mathbb{R}$ is the matrix

$$\begin{bmatrix} x & 2 & 0 & 3 \\ 1 & 2 & 3 & 3 \\ 1 & 0 & 1 & 1 \\ 1 & 1 & 1 & 3 \end{bmatrix}$$

invertible?

11.4 Given the matrices

$$A = \begin{bmatrix} 1 & 2 & 3 \\ -1 & 1 & 2 \\ 3 & -2 & 1 \end{bmatrix}, \quad B = \begin{bmatrix} -2 & 3 & 5 \\ 1 & -2 & 1 \\ 0 & -3 & -2 \end{bmatrix}$$

give MAPLE code to compute the characteristic polynomial of

$$A^3 - B^2 + AB(B^2 - A).$$

11.5 Write a MAPLE program to find the maximum value m of the determinant of all 3×3 matrices all of whose entries are 1 or 2. Modify your code to find all such 3×3 matrices with determinant m.

11.6 Let $A_n = [a_{ij}]_{n \times n}$ where

$$a_{ij} = \begin{cases} i^2 + j & \text{if } i < j; \\ j^2 & \text{otherwise,} \end{cases}$$

and let $B_n = [b_{ij}]_{n \times n}$ where $b_{ij} = i^2 + j^2$.

Write MAPLE code to compute the determinants of A_n and B_n for n from 1 to 10.

If $C_n = A_n - B_n$, write code to compute the determinant of C_n for n from 1 to 10.

Devise a program that produces lists ℓa and ℓc in which the i-th entry of ℓa is the determinant of A_i and the i-th entry of ℓc is the determinant of C_i.

Guess the determinant of C_n for arbitrary n and write a program to verify your conjecture for the first 20 values of n.

11.7 Let us compute the determinant of matrices whose (i,j)-th entry is given by a polynomial in i and j, for example:

```
> f1:=(i,j)->i^2+j^2:
Determinant(Matrix(6,6,f1));
```

MAPLE gives the answer 0.

Try some more matrices whose (i,j)-th entry is a polynomial in i and j, for example:

```
> f2:=(i,j)->i^6*j^2+j^3+12:
f3:=(i,j)->1+i^2*j-i^3*j^5:
f4:=(i,j)->(i+j^3)^3+(i^5-j^2-6*i*j^2)^2:
Determinant(Matrix(6,6,f2));
Determinant(Matrix(5,5,f3));
Determinant(Matrix(11,11,f4));
```

What do you observe?

There must be a theorem here. Can you prove it?

12

... but who were they?

This chapter is devoted to brief biographies of those mathematicians who were foremost in the development of the subject that we now know as Linear Algebra. We do not pretend that this list is exhaustive, but we include all who have been mentioned in the present book and in *Basic Linear Algebra*. Those included are:

Bessel, Friedrich Wilhelm;
Bezout, Étienne;
Cauchy, Augustin Louis;
Cayley, Arthur;
Fibonacci, Leonardo Pisano;
Fourier, Jean Baptiste Joseph;
Gram, Jorgen Pederson;
Hamilton, Sir William Rowan;
Hermite, Charles;
Hilbert, David;
Jordan, Marie Ennemond Camille;
Kronecker, Leopold;
Lagrange, Joseph-Louis;
Laplace, Pierre-Simon;
Lie, Marius Sophus;
Parseval des Chênes, Marc-Antoine;
Schmidt, Erhard;
Schwarz, Hermann Amandus;
Sylvester, James Joseph;
Toeplitz, Otto;
Vandermonde, Alexandre Théophile.

We refer the reader to the website

http://www-history.mcs.st-andrews.ac.uk/history/

for more detailed biographies of these and more than 1500 other mathematicians.

Bessel, Friedrich Wilhelm

Born: 22 July 1784 in Minden, Westphalia (now Germany).

Died: 17 March 1846 in Königsberg, Prussia (now Kaliningrad, Russia).

Wilhelm Bessel attended the Gymnasium in Minden for four years but he did not appear to be very talented. In January 1799, at the age of 14, he left school to become an apprentice to the commercial firm of Kulenkamp in Bremen. Bessel spent his evenings studying geography, Spanish, English and navigation. This led him to study astronomy and mathematics, and he began to make observations to determine longitude.

In 1804 Bessel wrote a paper on Halley's comet, calculating the orbit using data from observations made by Harriot in 1607. He sent his results to Heinrich Olbers who recognised at once the quality of Bessel's work. Olbers suggested further observations which resulted in a paper at the level required for a doctoral dissertation.

In 1806 Bessel accepted the post of assistant at the Lilienthal Observatory, a private observatory near Bremen. His brilliant work there was quickly recognised and both Leipzig and Greifswald universities offered him posts. However he declined both. In 1809, at the age of 26, Bessel was appointed director of Frederick William III of Prussia's new Königsberg Observatory and appointed professor of astronomy. He took up his new post on 10 May 1810. Here he undertook his monumental task of determining the positions and proper motions of over 50,000 stars.

Bessel used Bradley's data to give a reference system for the positions of stars and planets and also to determine the positions of stars. He determined the constants of precession, nutation and aberration winning him further honours, such as a prize from the Berlin Academy in 1815.

Bessel used parallax to determine the distance to 61 Cygni, announcing his result in 1838. His method of selecting this star was based on his own data for he chose the star which had the greatest proper motion of all the stars he had studied, correctly deducing that this would mean that the star was nearby. Since 61 Cygni is a relatively dim star it was a bold choice based on his correct understanding of the cause of the proper motions. Bessel announced his value of 0.314" which, given the diameter of the Earth's orbit, gave a distance of about 10 light years. The correct value of the parallax of 61 Cygni is 0.292".

Bessel also worked out a method of mathematical analysis involving what are now known as Bessel functions. He introduced these in 1817 in his study of a problem of Kepler of determining the motion of three bodies moving under mutual gravitation. These functions have become an indispensable tool in applied mathematics, physics and engineering.

Bessel functions appear as coefficients in the series expansion of the indirect perturbation of a planet, that is the motion caused by perturbations of the Sun. In 1824 he developed Bessel functions more fully in a study of planetary perturbations and published a treatise on them in Berlin. It was not the first time that special cases of the functions had appeared, for Jacob Bernoulli, Daniel Bernoulli, Euler and Lagrange had studied special cases of them earlier. In fact it was probably Lagrange's work on elliptical orbits that first led Bessel to work on such functions.

Bezout, Étienne

Born: 31 March 1730 in Nemours, France.
Died: 27 September 1783 in Basses-Loges (near Fontainbleau), France.

Family tradition almost demanded that Étienne Bezout follow in his father's and grandfather's footsteps as a magistrate in the town of Nemours. However once he had read Euler's works he wished to devote himself to mathematics. In 1756 he published a memoir on dynamics and later two papers investigating integration.

In 1758 Bezout was appointed an *adjoint* in mechanics of the Académie des Sciences, then he was appointed examiner of the Gardes de la Marine in 1763. One important task that he was given in this role was to compose a textbook specially designed for teaching mathematics to the students. Bezout is famed for the texbooks which came out of this assignment. The first was a four volume work which appeared in 1764-67. He was appointed to succeed Camus becoming examiner of the Corps d'Artillerie in 1768. He began work on another mathematics textbook and as a result he produced a six volume work which appeared between 1770 and 1782. This was a very successful textbook and for many years it was the book which students hoping to enter the École Polytechnique studied. His books came in for a certain amount of criticism for lack of rigour but, despite this, they were understood by those who needed to use mathematics and as a result were very popular and widely used. Their use spread beyond France for they were translated into English and used in North America.

Bezout is famed also for his work on algebra, in particular on equations. He was much occupied with his teaching duties after appointments in 1763 and he could devote relatively little time to research. He made a conscientious decision to restrict the range of his work so that he could produce worthwhile results in a narrow area. The way Bezout went about his research is interesting. He attacked quite general problems, but since an attack was usually beyond what could be achieved with the mathematical knowledge then available, he attacked special cases of the general problems which he could solve.

His first paper on the theory of equations examined how a single equation in a single unknown could be attacked by writing it as two equations in two unknowns. He made the simplifying assumption that one of the two equations was of a particularly simple form; for example he considered the case when one of the two equations had only two terms, a term of degree n and a constant term. Already this paper had introduced the topic to which Bezout would make his most important contributions, namely methods of elimination to produce from a set of simultaneous equations a single resultant equation in one of the unknowns.

He also did important work on the use of determinants in solving equations and as a result Sylvester, in 1853, called the determinant of the matrix of coefficients of the equations the 'Bezoutiant'. These and further papers published by Bezout in the theory of equations were gathered together in *Théorie générale des équations algébriques* which was published in 1779.

Cauchy, Augustin Louis

Born: 21 August 1789 in Paris, France.
Died: 23 May 1857 in Sceaux (near Paris), France.

In 1802 Augustin-Louis Cauchy entered the École Centrale du Panthéon where, following Lagrange's advice, he spent two years studying classical languages. He took the entrance examination for the École Polytechnique in 1805 and the examiner Biot placed him second. At the École Polytechnique he attended courses by Lacroix, de Prony and Hachette and was tutored in analysis by Ampère. In 1807 he graduated from the École Polytechnique and entered the École des Ponts et Chaussées, an engineering school. In 1810 Cauchy took up his first job in Cherbourg working on port facilities for Napoleon's English invasion fleet. In addition to a heavy workload he undertook mathematical research. Encouraged by Legendre and Malus, he submitted papers on polygons and polyhedra in 1812. Cauchy felt that he had to return to Paris if he was to make an impression with mathematical research. In September of 1812 he returned after becoming ill. It appears that the illness was not a physical one and was probably of a psychological nature resulting in severe depression.

Back in Paris, Cauchy investigated symmetric functions and in a 1812 paper he used 'determinant' in its modern sense. He reproved the earlier results on determinants and gave new results of his own on minors and adjoints. This 1812 paper gives the multiplication theorem for determinants for the first time. He was supposed to return to Cherbourg in February 1813 when he had recovered his health but this did not fit with his mathematical ambitions. He was allowed to work on the Ourcq Canal project rather than return to Cherbourg.

Cauchy wanted an academic career but failed with several applications. He obtained further sick leave and then, after political events prevented work on the Ourcq Canal, he was able to devote himself entirely to research for a couple of years.

In 1815 Cauchy lost out to Binet for a mechanics chair at the École Polytechnique, but then was appointed assistant professor of analysis there. In 1816 he won the Grand Prix of the French Academy of Sciences for a work on waves. He achieved real fame however when he submitted a paper to the Institute solving one of Fermat's claims on polygonal numbers made to Mersenne.

In 1817 when Biot left Paris for an expedition to the Shetland Islands in Scotland, Cauchy filled his post at the Collège de France. There he lectured on methods of integration which he had discovered, but not published, earlier. He was the first to make a rigorous study of the conditions for convergence of infinite series in addition to his rigorous definition of an integral. His text *Cours d'Analyse* in 1821 was designed for students at École Polytechnique and was concerned with developing the basic theorems of the calculus as rigorously as possible. He began a study of the calculus of residues in 1826, and in 1829 he defined for the first time a function of a complex variable.

In 1826 Cauchy, in the context of quadratic forms in n variables, used the term 'tableau'for the matrix of coefficients. He found its eigenvalues and gave results on diagonalisation of a matrix in the context of converting a form to the sum of squares. He also introduced the idea of similar matrices (but not the term) and showed that if two matrices are similar then they have the same characteristic equation. He also proved, again in the context of quadratic forms, that every real symmetric matrix is diagonalisable.

By 1830 Cauchy decided to take a break. He left Paris in September 1830, after the revolution in July, and spent a short time in Switzerland. Political events in France meant that Cauchy was now required to swear an oath of allegiance to the new regime and when he failed to return to Paris to do so he lost all his positions there. In 1831 Cauchy went to Turin and after some time there he accepted an offer from the King of Piedmont of a chair of theoretical physics. He taught in Turin from 1832.

In 1833 Cauchy went to Prague to tutor Charles X's grandson. He returned to Paris in 1838 and regained his position at the Academy but not his teaching positions because he still refused to take an oath of allegiance. He was elected to the Bureau des Longitudes in 1839 but, after refusing to swear the oath, was not appointed and could not attend meetings or receive a salary. When Louis Philippe was overthrown in 1848 Cauchy regained his university positions. He produced 789 mathematics papers, an incredible achievement.

Cayley, Arthur

Born: 16 August 1821 in Richmond, Surrey, England.
Died: 26 January 1895 in Cambridge, Cambridgeshire, England.

Arthur Cayley's father came from an English family, but lived in St Petersburg, Russia. It was there that Cayley spent the first eight years of his childhood before his parents returned to England. At school he showed great skill in numerical calculations and, after he moved to King's College School in 1835, his ability for advanced mathematics became clear. His mathematics teacher advised him to pursue mathematics rather than enter the family merchant business as his father wished.

In 1838 Cayley began his studies at Trinity College, Cambridge from where he graduated in 1842 as Senior Wrangler. While still an undergraduate he had three papers published. After winning a Fellowship he taught for four years at Cambridge, publishing twenty-eight papers in the Cambridge Mathematical Journal.

A Cambridge fellowship had a limited tenure so Cayley had to find a profession. He chose law and was admitted to the bar in 1849. He spent 14 years as a lawyer but, although very skilled in conveyancing (his legal speciality), he always considered it as a means to let him pursue mathematics. One of Cayley's friends was Sylvester who also worked at the courts of Lincoln's Inn in London and there they discussed deep mathematical questions throughout their working day. During his 14 years as a lawyer Cayley published about 250 mathematical papers.

In 1863 Cayley was appointed Sadleirian Professor of Pure Mathematics at Cambridge. This involved a very large decrease in income but he was happy to have the chance to devote himself entirely to mathematics. This he certainly did, publishing over 900 papers and notes covering nearly every aspect of mathematics. In 1841 he published the first English contribution to the theory of determinants. Sylvester introduced the idea of a matrix but Cayley quickly saw its significance and by 1853 published a note giving, for the first time, the inverse of a matrix.

In 1858 he published a memoir which is remarkable for containing the first abstract definition of a matrix. In it Cayley showed that the coefficient arrays studied earlier for quadratic forms and for linear transformations are special cases of this general concept. He defined a matrix algebra defining addition, multiplication, scalar multiplication and gave inverses explicitly in terms of determinants. He also proved that a 2×2 matrix satisfies its own characteristic equation. He stated that he had checked the result for 3×3 matrices, indicating its proof. The general case was later proved by Frobenius.

As early as 1849 Cayley set out his ideas on permutation groups. In 1854 he wrote two papers which are remarkable for their insight into abstract groups. Before this only permutation groups had been studied and even this was a radically new area, yet Cayley defined an abstract group. Significantly he realised that matrices and quaternions could form groups.

Cayley developed the theory of algebraic invariance, and his development of n-dimensional geometry has been applied in physics to the study of the space-time continuum. Cayley also suggested that euclidean and non-euclidean geometry are special types of a more general geometry and he united projective geometry and metrical geometry.

Fibonacci, Leonardo Pisano

Born: 1170 in (probably) Pisa (now in Italy).
Died: 1250 in (possibly) Pisa (now in Italy).

Fibonacci, or Leonardo Pisano to give him his correct name, was born in Italy but was educated in North Africa where his father held a diplomatic post. His father's job was to represent the merchants of the Republic of Pisa who were trading in Bugia, a Mediterranean port in northeastern Algeria.

Fibonacci travelled until around the year 1200 when he returned to Pisa. There he wrote a number of important texts which played an important role in reviving ancient mathematical skills and he made significant contributions of his own. Of his books *Liber abbaci* (1202), *Practica geometriae* (1220), *Flos* (1225), and *Liber quadratorum* have survived.

Fibonacci's work was made known to Frederick II, the Holy Roman Emperor, through the scholars at his court. These scholars suggested to Frederick that he meet Fibonacci when the court met in Pisa around 1225. At the court a number of problems were presented to Fibonacci as challenges. Fibonacci gave solutions to some in *Flos*, a copy of which he sent to Frederick II.

After 1228 only one reference to Fibonacci has been found. This was in a decree made in 1240 in which a salary was awarded to Fibonacci in recognition for the services that he had given to Pisa, advising on matters of accounting and teaching the people.

Liber abbaci was based on the arithmetic and algebra that Fibonacci had collected during his travels. The book, which went on to be widely copied and imitated, introduced the Hindu-Arabic place-valued decimal system and the use of Arabic numerals into Europe. Indeed, although mainly a book about the use of Arab numerals, which became known as algorism, simultaneous linear equations are also studied in this work. Many of the problems that Fibonacci considers in *Liber abbaci* were similar to those appearing in Arab sources.

The second section of *Liber abbaci* contains a large collection of problems aimed at merchants. They relate to the price of goods, how to calculate profit on transactions, how to convert between the various currencies in use in Mediterranean countries, and problems which had originated in China. A problem in the third section led to the introduction of the Fibonacci numbers and the Fibonacci sequence for which Fibonacci is best remembered today.

'A certain man put a pair of rabbits in a place surrounded on all sides by a wall. How many pairs of rabbits can be produced from that pair in a year if it is supposed that every month each pair begets a new pair which from the second month on becomes productive?'

The resulting sequence is 1, 1, 2, 3, 5, 8, 13, 21, 34, 55, ... (Fibonacci omitted the first term in *Liber abbaci*). This sequence, in which each number is the sum of the two preceding numbers, has proved extremely fruitful and appears in many different areas of mathematics and science.

There are also problems involving perfect numbers, the Chinese remainder theorem and summing arithmetic and geometric series.

Apart from his role in spreading the use of the Hindu-Arabic numerals and the rabbit problem, his contribution to mathematics has been largely overlooked.

Fourier, Jean Baptiste Joseph

Born: 21 March 1768 in Auxerre, Bourgogne, France.
Died: 16 May 1830 in Paris, France.

Joseph Fourier's first schooling was at Pallais's school, then he proceeded in 1780 to the École Royale Militaire of Auxerre where at first he showed talents for literature. By the age of thirteen, however, mathematics became his real interest and within a year he had completed a study of the six volumes of Bezout's *Cours de Mathematiques*.

In 1787 Fourier decided to train for the priesthood and entered the Benedictine abbey of St Benoit-sur-Loire. He was unsure if he was making the right decision in training for the priesthood and he corresponded with the professor of mathematics at Auxerre. He submitted a paper on algebra to Montucla in Paris and his letters suggest that he really wanted to make a major impact in mathematics.

Fourier did not take his religious vows. Having left St Benoit in 1789, he visited Paris and read a paper on algebraic equations at the Académie Royale des Sciences. In 1790 he became a teacher at the Benedictine college, École Royale Militaire of Auxerre, where he had studied. Up until this time there had been a conflict inside Fourier about whether he should follow a religious life or one of mathematical research. However in 1793 a third element was added to this conflict when he became involved in politics and joined the local Revolutionary Committee. He defended members of one political faction while in Orléans then returned to Auxerre to continue work on the revolutionary committee and to teach at the College. In July 1794 he was arrested, the charges relating to the Orléans incident, and imprisoned. Fourier feared that he would be put to the guillotine but, after Robespierre himself suffered that fate, political changes allowed him his freedom.

Later in 1794 Fourier was nominated to study at the École Normale in Paris. He was taught by Lagrange, Laplace and Monge then began teaching at the Collège de France where he again undertook research. He was appointed to the École Polytechnique but repercussions of his earlier arrest remained. Again he was arrested and imprisoned but fortunately soon released. In 1797 he succeeded Lagrange to the chair of analysis and mechanics at the École Polytechnique. He was renowned as an outstanding lecturer but he does not appear to have undertaken research around this time.

In 1798 Fourier joined Napoleon's army in its invasion of Egypt as a scientific adviser. He returned to France in 1801 with the remains of the expeditionary force and resumed his post as Professor of Analysis at the École Polytechnique. However Napoleon had other ideas about how Fourier might serve him and he was sent to Grenoble where his duties as Prefect were many and varied.

It was during his time in Grenoble that Fourier did his important mathematical work on the theory of heat. His work on the topic began around 1804 and by 1807 he had completed his important memoir which today is very highly regarded but at the time it caused controversy. Referees were unhappy with the work because of an objection, made by Lagrange and Laplace in 1808, to Fourier's expansions of functions as trigonometrical series, what we now call Fourier series. With a rather mixed report there was no move to publish the work.

When Napoleon escaped from Elba and marched into Grenoble, Fourier left in haste. However he was able to talk his way into favour with both sides and Napoleon made him Prefect of the Rhône. After Napoleon was defeated, Fourier returned to Paris. During his final eight years he resumed his mathematical research.

Gram , Jorgen Pedersen
Born: 27 June 1850 in Nustrup, Denmark.
Died: 29 April 1916 in Copenhagen, Denmark.

Jorgen Gram, the son of a farmer, entered the Ribe Katedralskole secondary school in 1862. He began his university education in 1868. In 1873 Gram graduated with a Master's degree in mathematics. He had published his first important mathematics paper before he graduated. This was a work on modern algebra which provided a simple and natural framework for invariant theory.

In 1875 Gram was appointed as an assistant in the Hafnia Insurance Company. Around the same time he began working on a mathematical model of forest management. His career in the Hafnia Insurance Company progressed well and his work for the company led him back into mathematical research. He began working on probability and numerical analysis, two topics whose practical applications in his day to day work in calculating insurance made their study important to him.

Gram's mathematical career was always a balance between pure mathematics and very practical applications of the subject. His work on probability and numerical analysis involved both the theory and its application to very practical situations. He published a paper on methods of least squares and for this work he was awarded the degree of Doctor of Science in 1879. Gram was led to study abstract problems in number theory. In 1884 he won the Gold Medal of the Videnskabernes Society for his paper *Investigations of the number of primes less than a given number.*

Although he continued to work for the Hafnia Insurance Company in more and more senior roles, Gram founded his own insurance company, the Skjold Insurance Company, in 1884. He was the director of this company from its founding until 1910. From 1895 until 1910 Gram was also an executive of the Hafnia Insurance Company, and from 1910 until his death in 1916 was Chairman of the Danish Insurance Council.

Despite not teaching mathematics in a university and as a consequence never having any students, Gram still managed to influence the next generation of Danish mathematicians in a very positive way. He often lectured in the Danish Mathematical Society, he was an editor of *Tidsskrift for Mathematik* from 1883 to 1889, and he also reviewed papers written in Danish for German mathematical journals.

Gram received honours for his mathematical contributions despite being essentially an amateur mathematician. The Videnskabernes Society had awarded him their Gold Medal in 1884 before he became a member, but in 1888 he was honoured with election to the Society. He frequently attended meetings of the Society and published in the Society's journals. For many years he was its treasurer.

Gram is best remembered for the Gram–Schmidt orthogonalisation process which constructs an orthogonal set from an independent one. He was not however the first to use this method. The process seems to be a result of Laplace and it was essentially used by Cauchy in 1836.

Gram met his death in a rather strange and very sad way. He was on his way to a meeting of the Videnskabernes Society when he was struck and killed by a bicycle. He was sixty-five years old when he met his death in this tragic accident.

Hamilton, Sir William Rowan

Born: 4 August 1805 in Dublin, Ireland.
Died: 2 September 1865 in Dublin, Ireland.

By the age of five, Hamilton had already learned Latin, Greek, and Hebrew. He was taught these subjects by his uncle with whom he lived with in Trim for many years. He soon mastered additional languages but a turning point came in his life at the age of 12 when he met the American Zerah Colburn, who could perform amazing mental arithmetical feats and with whom Hamilton joined in competitions of arithmetical ability. It appears that losing to Colburn sparked Hamilton's interest in mathematics.

Hamilton's introduction to mathematics came at the age of 13 when he studied Clairaut's *Algebra*. At age 15 he started studying the works of Newton and Laplace. In 1822 Hamilton found an error in Laplace's *Méchanique Céleste*. He entered Trinity College, Dublin at the age of 18 and in his first year he obtained an *optime* in Classics, a distinction awarded only once in 20 years.

Hamilton made remarkable progress for an undergraduate and before the end of 1824 submitted his first paper to the Royal Irish Academy on caustics. After an unhappy love affair he considered suicide, but instead turned to poetry, which he returned to throughout his life in times of anguish.

In his final year as an undergraduate he presented a memoir *Theory of Systems of Rays* to the Royal Irish Academy. In 1827 he was appointed Professor of Astronomy at Trinity College while he was still an undergraduate aged twenty-one years. This appointment brought a great deal of controversy as Hamilton did not have much experience in observing.

Before beginning his duties in this prestigious position, Hamilton toured England and Scotland (from where the family name originates). He met the poet Wordsworth and they became friends. One of Hamilton's sisters also wrote poetry and when

Wordsworth came to visit them he indicated a preference for her poems rather than his. The two men had long debates over poetry versus science, the culmination of which was that Worsdworth had to tell Hamilton that his talents lay in the latter:

You send me showers of verses which I receive with much pleasure .... yet have we fears that this employment may seduce you from the path of science .... Again I do venture to submit to your consideration whether the poetical parts of your nature would not find a field more favourable to their nature in the regions of prose, not because those regions are humbler, but because they may be gracefully and profitably trod, with footsteps less careful and in measures less elaborate.

In 1832 Hamilton published a third supplement to *Theory of Systems of Rays* which is essentially a treatise on the characteristic function applied to optics. Near the end of the work he applied the characteristic function to study Fresnel's wave surface. From this he predicted conical refraction and asked the Professor of Physics at Trinity College to try to verify his theoretical prediction experimentally. This happened two months later and brought him great fame.

In 1833 Hamilton read a paper to the Royal Irish Academy expressing complex numbers as algebraic couples, or ordered pairs of real numbers. After the discovery of algebraic couples, he tried to extend the theory to triplets, and this became an obsession that plagued him for many years.

On 16 October 1843 (a Monday) Hamilton was walking in along the Royal Canal with his wife to preside at a Council meeting of the Royal Irish Academy. Although his wife talked to him now and again Hamilton hardly heard, for the discovery of the quaternions, the first noncommutative algebra to be studied, was taking shape in his mind. He could not resist the impulse to carve the formulae for the quaternions

$$i^2 = j^2 = k^2 = ijk = -1$$

in the stone of Brougham Bridge as he passed it. Hamilton felt this discovery would revolutionise mathematical physics and spent the rest of his life working on quaternions. He published *Lectures on Quaternions* in 1853 but he soon realised that it was not a good book from which to learn the theory. Perhaps Hamilton's lack of skill as a teacher showed up in this work.

Determined to produce a book of some lasting quality, Hamilton began to write *Elements of Quaternions* which he estimated would be 400 pages long and take two years to write. The title suggests that he modelled his work on Euclid's *Elements*. The book was twice the intended length and took seven years to write, the final chapter being incomplete when he died. It was published with a preface by his son.

Hamilton's personal life was not a happy one, and was exacerbated by the suicide of his colleague James MacCullagh at Trinity College. His ever-increasing dependancy on alcoholic stimulants brought little respite. He died from a severe attack of gout, shortly after receiving the news that he had been elected the first Foreign Member of the National Academy of Sciences of the United States of America.

Hermite, Charles

Born: 24 December 1822 in Dieuze, Lorraine, France.

Died: 14 January 1901 in Paris, France.

Hermite's parents did not take much personal interest in their children's education, although they did provide them with good schooling. He was something of a worry to his parents for he had a defect in his right foot which meant that he moved around only with difficulty.

He attended the Collège de Nancy, then went to Paris where he attended the Collège Henri. In 1840-41 he studied at the Collège Louis-le-Grand where some fifteen years earlier Galois had studied. He preferred to read papers by Euler, Gauss and Lagrange rather than work for his formal examinations but he showed remarkable research ability, publishing two papers while at Louis-le-Grand.

Hermite wanted to study at the École Polytechnique and he took a year preparing for the examinations being tutored by Catalan in 1841-42. He passed but only attained sixty-eighth place in the ordered list. After one year at the École Polytechnique he was refused the right to continue his studies because of his disability. The decision was reversed but strict conditions which Hermite did not find acceptable were imposed and he decided not to graduate from the École Polytechnique.

Hermite made friends with important mathematicians at this time and frequently visited Joseph Bertrand, whose sister Louise he later married. He exchanged letters with Jacobi which show that Hermite had discovered some differential equations satisfied by theta-functions and he was using Fourier series to study them. After spending five years working towards his degree he passed in 1847. In the following year he was appointed to the École Polytechnique, the institution which had tried to prevent him continuing his studies some four years earlier.

Hermite made important contributions to number theory and algebra, orthogonal polynomials, and elliptic functions. He discovered his most significant mathematical results over the ten years following his appointment to the École Polytechnique. One topic on which Hermite worked was the theory of quadratic forms. This led him to study invariant theory and he found a reciprocity law relating to binary forms. With his understanding of quadratic forms and invariant theory he created a theory of transformations in 1855.

The next mathematical result by Hermite which we must mention is one for which he is rightly famous. Although an algebraic equation of the fifth degree cannot be solved in radicals, Hermite showed in 1858 that an algebraic equation of the fifth degree could be solved using elliptic functions. He applied these results to number theory, in particular to class number relations of quadratic forms.

The year 1869 saw Hermite become a professor of analysis when he succeeded Duhamel both at the École Polytechnique and at the Sorbonne. The 1870s saw Hermite return to problems concerning approximation and interpolation in which he had been interested earlier in his career. In 1873 he published the first proof that e is a transcendental number. He is now best known for Hermite polynomials, Hermite's differential equation, Hermite's formula of interpolation and hermitian matrices. He resigned his chair at the École Polytechnique in 1876 but continued to hold the chair at the Sorbonne until he retired in 1897.

Hilbert, David

Born: 23 January 1862 in Königsberg, Prussia (now Kaliningrad, Russia).
Died: 14 February 1943 in Göttingen, Germany.

David Hilbert attended the gymnasium in his home town of Königsberg. After graduating from this school, he entered the University of Königsberg. There he went on to study under Lindemann for his doctorate which he received in 1885. Hilbert was a member of staff at Königsberg from 1886 to 1895, being a Privatdozent until 1892, then as Extraordinary Professor for one year before being appointed a full professor in 1893. In 1895 Klein appointed Hilbert to the chair of mathematics at the University of Göttingen, where he continued to teach for the rest of his career.

Hilbert's eminent position in the world of mathematics after 1900 meant that other institutions would have liked to tempt him to leave Göttingen and, in 1902, the University of Berlin offered him a chair. Hilbert turned down the offer, but only after he had used it to bargain with Göttingen and persuade them to set up a new chair to bring his friend Minkowski there.

Hilbert contributed to many branches of mathematics, including invariant theory, algebraic number fields, functional analysis, integral equations, mathematical physics, and the calculus of variations. His first work was on invariant theory and, in 1888, he proved his famous Basis Theorem. He discovered a completely new approach which proved the theorem for any number of variables in an entirely abstract way. Although this proved that a finite basis existed, his methods did not construct such a basis. Hilbert's revolutionary approach was difficult for others to appreciate and he had to argue his case strongly before Klein agreed that the work could be published.

In 1893 while still at Königsberg Hilbert began his *Zahlbericht* on algebraic number theory. The German Mathematical Society had requested this major report three years after the Society was created in 1890. The ideas of the present day subject of class field theory are all contained in this work.

Hilbert's work in geometry had the greatest influence in that area after Euclid. A systematic study of the axioms of Euclidean geometry led Hilbert to propose 21 such axioms and he analysed their significance. He published *Grundlagen der Geometrie* in 1899, putting geometry in a formal axiomatic setting. The book continued to appear in new editions and was a major influence in promoting the axiomatic approach to mathematics which has been one of the major characteristics of the subject throughout the 20th century.

In his famous speech, delivered at the Second International Congress of Mathematicians in Paris, his 23 problems challenged (and still today challenge) mathematicians to solve fundamental questions.

The ideas of a Hilbert space came out of his work on integral equations. It was in this context that he introduced the words 'eigenvalue'and 'eigenfunction'. In his work with his student Schmidt, the ideas of abstract infinite dimensional spaces evolved around 1904. However, the fully axiomatic approach did not appear until Banach's 1920 doctoral dissertation.

In 1930 Hilbert retired and the city of Königsberg made him an honorary citizen. His address ended with words showing his devotion to solving mathematical problems: 'We must know, we shall know'.

Jordan, Marie Ennemond Camille

Born: 5 January 1838 in Lyon, France.
Died: 22 January 1922 in Milan, Italy.

Camille Jordan entered the École Polytechnique to study mathematics in 1855. His doctoral thesis was examined in January 1861, after which he worked as an engineer, first at Privas, then at Chalon-sur-Saône, and finally in Paris.

From 1873 he was an examiner at the École Polytechnique where he became professor of analysis in November 1876. He was also a professor at the Collège de France from 1883 although until 1885 he was at least theoretically still an engineer by profession. It is significant, however, that he found more time to undertake research when he was an engineer.

Jordan was a mathematician who worked in a wide variety of different areas essentially contributing to every mathematical topic which was studied at that time: finite groups, linear and multilinear algebra, the theory of numbers, the topology of polyhedra, differential equations, and mechanics.

Topology (called *analysis situs* at that time) played a major role in some of his first publications which were a combinatorial approach to symmetries. He introduced important topological concepts in 1866 building on his knowledge of Riemann's work. He introduced the notion of homotopy of paths looking at the deformation of paths one into the other. He defined a homotopy group of a surface without explicitly using group terminology.

Jordan was particularly interested in the theory of permutation groups. He introduced the concept of a composition series and proved the Jordan–Hölder theorem. Jordan clearly saw classification as an aim of the subject, even if it was not one which might ever be solved. He made some remarkable contributions to how such a classification might proceed setting up a recursive method to determine all soluble groups of a given order.

His work on group theory done between 1860 and 1870 was written up into a major text which he published in 1870. This treatise gave a comprehensive study of

Galois theory as well as providing the first ever group theory book. The treatise contains the 'Jordan normal form' theorem for matrices, not over the complex numbers but over a finite field.

Jordan's use of the group concept in geometry in 1869 was motivated by studies of crystal structure. He went on to produce further results of fundamental importance. He studied primitive permutation groups and, generalising a result of Fuchs on linear differential equations, he was led to study the finite subgroups of the general linear group of $n \times n$ matrices over the complex numbers. Although there are infinite families of such finite subgroups, Jordan found that they were of a very specific group theoretic structure which he was able to describe.

Another generalisation, this time of work by Hermite on quadratic forms with integral coefficients, led Jordan to consider the special linear group of $n \times n$ matrices of determinant 1 over the complex numbers acting on the vector space of complex polynomials of degree m in n indeterminates.

Jordan is best remembered today among analysts and topologists for his proof that a simply closed curve divides a plane into exactly two regions, now called the Jordan curve theorem. Among Jordan's many contributions to analysis should also be mentioned his generalisation of the criteria for the convergence of a Fourier series.

Kronecker, Leopold

Born: 7 December 1823 in Liegnitz, Prussia (now Legnica, Poland).
Died: 29 December 1891 in Berlin, Germany .

Leopold Kronecker was taught mathematics at Liegnitz Gymnasium by Kummer who stimulated his interest in mathematics. Kummer immediately recognised Kronecker's talent for mathematics and he took him well beyond what would be expected at school, encouraging him to undertake research.

Kronecker became a student at Berlin University in 1841 and there he studied under Dirichlet and Steiner. He did not restrict himself to mathematics, however, for he studied other topics such as astronomy, meteorology, chemistry and philosophy. After spending the summer of 1843 at the University of Bonn, where he went because of his interest in astronomy rather than mathematics, he travelled to the University of Breslau for the winter semester of 1843-44. He went there because he wanted to study mathematics with his old school teacher Kummer who had been appointed to a chair at Breslau in 1842. He spent a year there before returning to Berlin for the winter semester of 1844-45 where he worked on his doctoral thesis on algebraic number theory under Dirichlet's supervision.

Just as it looked as if he would embark on an academic career, Kronecker left Berlin to manage the banking business of his mother's brother. He also managed a family estate but still found the time to continue working on mathematics, although he did this entirely for his own enjoyment.

In 1855 Kronecker came to Berlin but not to a university appointment. He did not lecture at this time but was remarkably active in research, publishing a large number of works in quick succession. These were on number theory, elliptic functions and algebra, but, more importantly, he explored the interconnections between these topics. Kronecker was elected to the Berlin Academy which gave him the right to lecture at the University and this he did beginning in 1862. The topics on which he lectured were very much related to his research: number theory, the theory of equations, the theory of determinants, and the theory of integrals.

As we have already indicated, Kronecker's primary contributions were in the theory of equations and higher algebra, with his major contributions in elliptic functions, the theory of algebraic equations, and the theory of algebraic numbers. However the topics he studied were restricted by the fact that he believed in the reduction of all mathematics to arguments involving only a finite number of steps. Kronecker is well known for his remark: 'God created the integers, all else is the work of man'.

He was the first to doubt the significance of non-constructive existence proofs. It appears that, from the early 1870s, he was opposed to the use of irrational numbers, upper and lower limits, and the Bolzano–Weierstrass theorem, because of their non-constructive nature.

Kronecker had no official position at Berlin until Kummer retired in 1883 when he was appointed to the chair. By 1888 Weierstrass felt that he could no longer work with Kronecker and decided to go to Switzerland, but then, realising that Kronecker would be in a strong position to influence the choice of his successor, he decided to remain in Berlin!

Lagrange, Joseph-Louis

Born: 25 January 1736 in Turin, Sardinia-Piedmont (now Italy).
Died: 10 April 1813 in Paris, France.

Joseph-Louis Lagrange studied at the College of Turin where his favourite subject was classical Latin. At first he had no enthusiasm for mathematics, finding Greek geometry rather dull.

Lagrange decided to make mathematics his career after reading Halley's 1693 work on the use of algebra in optics. In 1754, despite having no training in advanced mathematics, he published his first mathematical work. It was no masterpiece and suffered because Lagrange was working without the advice of a mathematical supervisor.

Next he studied the tautochrone, the curve on which a particle will always arrive at a fixed point in the same time independent of its initial position. By the end of 1754 his discoveries on this topic made a major contribution to the new subject of the calculus of variations. Lagrange was appointed professor of mathematics at the Royal Artillery School in Turin in 1755. It was well deserved for he had already shown great originality and depth of thinking.

After Euler proposed him for the Berlin Academy, Lagrange was elected in 1756. The following year he was a founding member of a scientific society, later the Royal Academy of Science, in Turin. This new Society began publishing a scientific journal *Mélanges de Turin* with Lagrange a major contributor to the first few volumes with papers on the calculus of variations, calculus of probabilities, foundations of dynamics, propagation of sound, and the integration of differential equations with various applications to topics such as fluid mechanics.

The libration of the Moon was announced as the topic for the 1764 prize competition of the Académie des Sciences. Lagrange entered the competition, sending his entry to Paris in 1763. It arrived there not long before Lagrange himself but he took ill shortly after. Returning to Turin in early 1765, he submitted an entry for the 1766 prize on the orbits of the moons of Jupiter.

D'Alembert, who had visited the Berlin Academy and was friendly with Frederick II of Prussia, arranged for Lagrange to be offered a position in the Berlin Academy. Despite Lagrange's poor position in Turin, he at first refused but later accepted. He succeeded Euler as Director of Mathematics at the Berlin Academy of Science in 1766.

For 20 years Lagrange worked at Berlin, producing a stream of top quality papers and regularly winning the prize from the Académie des Sciences in Paris. He shared the 1772 prize on the three body problem with Euler, won the prize for 1774, another one on the motion of the moon, and he won the 1780 prize on perturbations of the orbits of comets by the planets. His work covered many topics: astronomy, the stability of the solar system, mechanics, dynamics, fluid mechanics, probability, and the foundations of the calculus. He also worked on number theory proving in 1770 that every positive integer is the sum of four squares. In 1771 he proved Wilson's theorem that n is prime if and only if $(n-1)! + 1$ is divisible by n. His 1770 paper *Réflexions sur la résolution algébrique des équations* made a fundamental investigation of why equations of degrees up to 4 could be solved by radicals. For the first time the roots of an equation are considered as abstract quantities rather than having numerical values. He studied permutations of the roots and, despite not composing permutations in the paper, it was a first step in the development of group theory. Lagrange, in a paper of 1773, studied identities for 3×3 functional determinants. However this comment is made with hindsight since Lagrange himself saw no connection between his work and that of Laplace and Vandermonde. This paper contains the volume interpretation of a determinant for the first time. Lagrange showed that the tetrahedron formed by $O(0,0,0)$ and the three points $M(x, y, z), M'(x', y', z'), M''(x'', y'', z'')$ has volume

$$\tfrac{1}{6}[z(x'y'' - y'x'') + z'(yx'' - xy'') + z''(xy' - yx')].$$

In 1787 Lagrange left Berlin for the Académie des Sciences in Paris, where he remained for the rest of his career. His *Mécanique Analytique*, written in Berlin, was published in 1788. It summarised the field of mechanics and transformed the topic into a branch of mathematical analysis.

Lagrange was the first professor of analysis at the École Polytechnique, appointed for its opening in 1794. In 1795 the École Normale was founded with the aim of training school teachers and Lagrange taught courses on elementary mathematics there.

Napoleon named Lagrange to the Legion of Honour and Count of the Empire in 1808. He was awarded the Grand Croix de l'Ordre Impérial de la Réunion a week before he died.

Laplace, Pierre-Simon

Born: 23 March 1749 in Beaumont-en-Auge, Normandy, France.
Died: 5 March 1827 in Paris, France.

Pierre-Simon Laplace attended a Benedictine priory school in Beaumont-en-Auge. His father expected him to make a career in the Church and, at the age of 16, he entered Caen University and enrolled in theology. However Laplace soon discovered his love of mathematics.

He left Caen without taking his degree, and went to Paris with a letter of introduction to d'Alembert from his teacher at Caen. Although only 19 years old he quickly impressed d'Alembert who tried to find him a position so he could support himself. Laplace was soon appointed as professor of mathematics at the École Militaire and he began producing a steady stream of remarkable mathematical papers.

In 1773 he was elected to the Académie des Sciences. By this time he had read 13 papers to the Academy in less than three years on: difference equations, differential equations, mathematical astronomy and the theory of probability. In 1772 Laplace claimed that the methods introduced by Cramer and Bezout were impractical and, in a paper where he studied the orbits of the inner planets, he discussed the solution of systems of linear equations using determinants. Rather surprisingly Laplace used the word 'resultant' for what we now call the determinant; surprising since it is the same word used by Leibniz yet Laplace must have been unaware of Leibniz's work. Laplace gave the expansion of a determinant which is now named after him.

His papers in the 1780s would make Laplace one of the most important and influential scientists that the world has seen. It was not achieved, however, with good relationships with his colleagues. Although d'Alembert had been proud to have considered Laplace as his protégé, he certainly began to feel that Laplace was rapidly making much of his own life's work obsolete and this did nothing to improve relations.

In 1784 Laplace was appointed as examiner at the Royal Artillery Corps, and in this role in 1785, he examined and passed the 16-year-old Napoleon Bonaparte. Laplace was made a member of the committee of the Académie des Sciences to standardise weights and measures in 1790. This committee worked on the metric system and advocated a decimal base. In 1793 the Reign of Terror commenced and the Académie des Sciences, along with the other learned societies, was suppressed on 8 August. The weights and measures commission was the only one allowed to continue but soon Laplace was thrown off the commission. With his wife and two children, he left Paris, not returning until after July 1794.

Laplace presented his famous nebular hypothesis in 1796 which viewed the solar system as originating from the contracting and cooling of a large, flattened, and slowly rotating cloud of incandescent gas. He put all his work in this area into his great work the *Traité du Mécanique Céleste* published in 5 volumes, the first two in 1799.

Under Napoleon, Laplace was a member, then chancellor, of the Senate, and admitted to the Legion of Honour in 1805. However Napoleon, in his memoirs written on St Hélène, says he removed Laplace from the office of Minister of the Interior after only six weeks '... because he brought the spirit of the infinitely small into the government'. Laplace became Count of the Empire in 1806 and he was named a marquis in 1817 after the restoration of the Bourbons.

The first edition of Laplace's *Théorie Analytique des Probabilités* was published in 1812. In 1814 Laplace supported the restoration of the Bourbon monarchy and cast his vote in the Senate against Napoleon. The Hundred Days were an embarrassment to him the following year and he conveniently left Paris for the critical period. After this he remained a supporter of the Bourbon monarchy and became unpopular in political circles. When he refused to sign the document of the French Academy supporting freedom of the press in 1826, he lost his remaining friends in politics.

Lie, Marius Sophus

Born: 17 December 1842 in Nordfjordeide, Norway.
Died: 18 February 1899 in Kristiania (now Oslo), Norway .

Sophus Lie first attended school in the town of Moss then, in 1857, he entered Nissen's Private Latin School in Christiania. He decided to take up a military career, but his eyesight was not sufficiently good so he gave up the idea and entered the University of Christiania.

At university Lie studied a broad science course, and he attended lectures by Sylow in 1862. Surprisingly he graduated in 1865 without having shown any great ability for mathematics, or any real liking for it. He considered a career in astronomy or botany or zoology or physics and in general became rather confused. From 1866 he began to read more and more mathematics and his interests steadily turned in that direction.

In 1867 Lie had his first brilliant new mathematical idea. It came to him in the middle of the night and, filled with excitement, he rushed to tell a friend. Lie wrote a short mathematical paper in 1869, which he published at his own expense, based on this inspiration. He wrote up a more detailed exposition, but the world of mathematics was too cautious to quickly accept Lie's revolutionary notions. The Academy of Science in Christiania was reluctant to publish his work, and at this stage Lie began to despair that he would become accepted in the mathematical world. The breakthrough came later in 1869 when Crelle's Journal accepted his paper.

Setting off near the end of 1869, Lie visited Göttingen and then Berlin. In Berlin he met Kronecker, Kummer and Weierstrass but he was not attracted to Weierstrass's mathematics. However he met Felix Klein and the two instantly found common ground in mathematics. In the spring of 1870 Lie and Klein were together again in Paris where Camille Jordan succeeded in a way that Sylow did not, for Jordan made Lie realise how important group theory was for the study of geometry. Lie's new ideas later appeared in his work on transformation groups.

While Lie and Klein thought deeply about mathematics in Paris, the political situation between France and Prussia deteriorated. Lie decided to remain but became anxious as the German offensive met with an ineffective French reply. When the German army trapped part of the French army in Metz, Lie decided it was time to leave and he decided to hike to Italy. He reached Fontainebleau but there he was arrested as a German spy, his mathematics notes being assumed to be top secret coded messages. Only after the intervention of Darboux was he released from prison.

In 1871 Lie became an assistant at Christiania and also taught at Nissen's Private Latin School. He submitted a dissertation *On a class of geometric transformations* for his doctorate which was duly awarded in July 1872. Lie started examining partial differential equations, hoping that he could find a theory which was analogous to the Galois theory of equations. This led to combining the transformations in a way that Lie called an infinitesimal group, what is today called a Lie algebra. When Klein left the chair at Leipzig in 1886, Lie was appointed to succeed him. In Leipzig, his life was rather different from that in Christiania. He was now in the mainstream of mathematics and students came from many countries to study under him. However his health was already deteriorating when he returned to a chair in Christiania in 1898, and he died of pernicious anaemia soon after taking up the post.

Parseval des Chênes, Marc-Antoine

   Born: 27 April 1755 in Rosières-aux-Saline, France.
   Died: 16 August 1836 in Paris, France .

Very little is known of Antoine Parseval's life. We know that he was born into a family of high standing in France and he describes himself as a squire, certainly suggesting that his family were wealthy land-owners. One of the few pieces of information which exists is that he married Ursule Guerillot in 1795. The marriage certainly did not last long and the pair were soon divorced.

The starting point of the historical events which were to play a major role in Parseval' life was the storming of the Bastille on 14 July 1789. Parseval, perhaps not surprisingly since he was of noble birth, was a royalist so for him the increasing problems for the monarchy meant that his life was more and more in danger. In 1792 Louis 16th gave up attempts at a compromise with opponents of the monarchy and tried to flee from France. He did not make it but was arrested and brought back to Paris. It was a time of great danger for royalist supporters and indeed it proved so for Parseval who was imprisoned in 1792.

Following the execution of the King on 21 January 1793 there followed a reign of terror with many political trials. Parseval, despite being freed from prison, remained in fear of his life. Napoleon became 1st Consul in 1800 and then Emperor in 1804. Parseval should have kept his head down if he wanted to avoid trouble but it was a time when it was almost impossible not to get drawn into the political events. Parseval published poetry against Napoleon's regime and, not surprisingly, had to flee from France when Napoleon ordered his arrest. He was successful in avoiding arrest and was able to return to Paris.

Parseval had only five publications, all presented to the Académie des Sciences between 5 May 1798 and 7 May 1804. It was the second of these, dated 5 April 1799, which contains the result known today as Parseval's theorem. Today this theorem is seen in the context of Fourier series, and often also in more abstract settings which are quite far removed from Parseval's original ideas. The original theorem was concerned with summing infinite series and had some restrictive conditions which Parseval removed in a note appended to an 1801 paper.

The theorem was not published until his five papers were all published by the Académie des Sciences in 1806. Before that it was known by members of the Academy and appeared in works by Lacroix and Poisson before Parseval's papers were printed.

Parseval was never honoured with election to the Académie des Sciences. He was proposed on five separate occasions, namely in 1796, 1799, 1802, 1813 and 1828. He was never particularly close although he did come third in 1799, the year that Lacroix was elected. It would not be unfair to say that Parseval has fared well in having a well known result, which is quite far removed from his contribution, named after him. However he remains a somewhat shadowy figure and it is hoped that research will one day provide a somewhat better understanding of his life and achievements.

Schmidt, Erhard

Born: 13 January 1876 in Dorpat, Russia (now Tartu, Estonia).
Died: 6 December 1959 in Berlin, Germany.

Erhard Schmidt attended his local university in Dorpat before going to Berlin where he studied with Schwarz. His doctorate was obtained from the University of Göttingen in 1905 under Hilbert's supervision. His doctoral dissertation concerned integral equations. The main ideas of this thesis appeared in a paper in 1907 of which we give more details below. After obtaining his doctorate he went to Bonn where he was awarded his *habilitation* in 1906. Schmidt went on to hold positions in Zurich, Erlangen and Breslau before he was appointed to a professorship at the University of Berlin in 1917.

Clearly Schmidt's organisational abilities were recognised outside mathematics for he was appointed Dean for the academic year 1921–22 and the Vice-Chancellor of the University of Berlin during the years 1929–30. The inaugural address that he gave when taking up the post of Vice-Chancellor was entitled *On certainty in mathematics*.

After the end of World War II Schmidt was appointed as Director of the Mathematics Research Institute of the German Academy of Science, remaining in that role until 1958. By that time he had retired from his chair, which he did in 1950. Another role which he took on after the end of the war was as the first editor of *Mathematische Nachrichten*, a journal he co-founded in 1948.

Schmidt's main interest was in integral equations and Hilbert space. He took various ideas of Hilbert on integral equations and combined these into the concept of a Hilbert space around 1905. Hilbert had shown that integral equations with symmetric kernels had real eigenvalues, and the solutions corresponding to these eigenvalues he called eigenfunctions. He also expanded functions related to the integral of the kernel function as an infinite series in a set of orthonormal eigenfunctions.

Schmidt published a two part paper on integral equations in 1907 in which he reproved Hilbert's results in a simpler fashion, and also with less restrictions. In this paper he gave what is now called the Gram–Schmidt orthonormalisation process for constructing an orthonormal set of functions from a linearly independent set. He then went on to consider the case where the kernel is not symmetric and showed that in that case the eigenfunctions associated with a given eigenvalue occurred in adjoint pairs. We should note, however, that Laplace presented the Gram–Schmidt process before either Gram or Schmidt.

In 1908 Schmidt published an important paper on infinitely many equations in infinitely many unknowns, introducing various geometric notations and terms which are still in use for describing spaces of functions and also in inner product spaces. Schmidt's ideas were to lead to the geometry of Hilbert spaces and he must certainly be considered as a founder of functional analysis.

Schmidt defined a space H whose elements are square summable sequences of complex numbers, defining a norm in terms of the inner product. He also defined orthogonal elements, showing that a set consisting of pair-wise orthogonal elements is linearly independent. Again he gave the Gram–Schmidt orthonormalisation process in this setting. He also studied projections and spectral resolutions. What are today called Hilbert–Schmidt operators also appear in this 1908 paper.

Schwarz, Hermann Amandus

Born: 25 January 1843 in Hermsdorf, Silesia (now Sobiecin, Poland).

Died: 30 November 1921 in Berlin, Germany.

Hermann Schwarz studied at the Gymnasium in Dortmund where his favourite subject was chemistry. He intended to take a degree in chemistry and he entered the Gewerbeinstitut, later called the Technical University of Berlin, with this aim. However, it wasn't long before Kummer and Weierstrass had influenced him to change to mathematics. Through Karl Pohlke, one of his teachers, Schwarz became interested in geometry. He attended Weierstrass's lectures on the integral calculus and his interest in geometry was soon combined with Weierstrass's ideas of analysis. He continued to study in Berlin, being supervised by Weierstrass, until 1864 when he was awarded his doctorate. His doctoral dissertation was examined by Kummer, his future father-in-law.

While in Berlin, Schwarz worked on minimal surfaces, a characteristic problem of the calculus of variations. He had made an important contribution in 1865 when he discovered what is now known as the Schwarz minimal surface. This minimal surface has a boundary consisting of four edges of a regular tetrahedron.

Schwarz continued studying in Berlin for his teacher's training qualification and completed this by 1867 when was appointed as a Privatdozent to the University of Halle. In 1869 he became professor of mathematics at the Eidgenössische Technische Hochschule in Zurich then, in 1875, was appointed to the chair of mathematics at Göttingen University.

Schwarz accepted the professorship in Berlin in 1892, but after this the university lost its leading role for mathematics. That this happened shouldn't have come as a complete surprise to those making the appointment, for Schwarz had published his *Complete Works* in 1890, two years earlier.

Schwarz continued teaching at Berlin until 1918. Outside mathematics he was the captain of the local Voluntary Fire Brigade and, more surprisingly, he assisted the stationmaster at the local railway station by closing the doors of the trains!

One important area which Schwarz worked on was that of conformal mappings. His most important work was a *Festschrift* for Weierstrass's 70th birthday in which he answered the question of whether a given minimal surface really yields a minimal area. An idea in this work, where he constructed a function using successive approximations, led Émile Picard to his existence proof for solutions of differential equations. It also contains the inequality for integrals now known as the 'Schwarz inequality'.

The fact that Schwarz should have come up with a special case of the general result now known as the Cauchy–Schwarz inequality (or the Cauchy–Bunyakovsky–Schwarz inequality) is not surprising, for much of his work is characterised by looking at rather specific and narrow problems but solving them using methods of great generality which have since found widespread applications. That he found such general methods says much for his great intuition which was perhaps based on a deep feeling for geometry. The form in which the inequality is usually presented today with its standard modern proof was first given by Weyl in 1918.

Schwarz's demeanour has been described as naive, dramatic, and coarse. In spite of giving the impression of self-confidence, he was, in fact, rather insecure and besides, not efficient in business matters.

Sylvester, James Joseph

Born: 3 September 1814 in London, England.
Died: 15 March 1897 in London, England.

James Joseph Sylvester attended two primary schools in London, then his secondary schooling was at the Royal Institution in Liverpool. In 1833 he became a student at St John's College, Cambridge. At this time it was necessary for a student to sign a religious oath to the Church of England before graduating and Sylvester, being Jewish, refused to take the oath. For the same reason he was not eligible for a Smith's prize nor for a Fellowship.

From 1838 Sylvester taught physics for three years at the University of London, one of the few places which did not bar him because of his religion. At the age of 27 he was appointed to a chair in the University of Virginia but he resigned after a few months. A student who had been reading a newspaper in one of his lectures insulted him and Sylvester struck him with a sword stick. The student collapsed in shock and Sylvester, believing (wrongly) that he had killed him, fled and boarded the first ship back to England.

On his return Sylvester worked as an actuary and lawyer but gave mathematics tuition. Cayley was also a lawyer and they worked together at the courts of Lincoln's Inn in London. Sylvester tried hard to return to being a professional mathematician and, after several failed applications, became professor of mathematics at Woolwich.

Sylvester did important work on matrix theory, discussing the topic with Cayley while they were at the courts of Lincoln's Inn. In particular he invented the term 'matrix' in 1850, defining a matrix to be an oblong arrangement of terms which he saw as something which led to various determinants from square arrays contained within it. He used matrix theory to study higher dimensional geometry. In 1851 he discovered the discriminant of a cubic equation and first used the name 'discriminant' for such expressions in higher order equations. He defined the nullity of a square matrix in 1884.

Being at a military academy Sylvester had to retire at age 55. His only book at this time was on poetry. Clearly Sylvester was proud of this work for he sometimes signed himself 'J. J. Sylvester, author of *The Laws of Verse*'. For three years Sylvester did no mathematical research but then Chebyshev visited London and the two discussed mechanical linkages which can be used to draw straight lines. Sylvester then lectured on this topic at evening lectures at the Royal Institution entitled *On recent discoveries in mechanical conversion of motion.*

In 1877 Sylvester accepted a chair at Johns Hopkins University in the United States and in 1878 he founded the *American Journal of Mathematics*, the first mathematical journal in the USA.

In 1883 Sylvester, although 68 years old at this time, was appointed to the Savilian chair of Geometry at Oxford. However he liked to lecture only on his own research and this was not well liked at Oxford. In 1892 Sylvester, by this time partially blind and suffering from loss of memory, returned to London where he spent his last years at the Athenaeum Club.

Toeplitz, Otto

Born: 1 August 1881 in Breslau, Germany (now Wroclaw, Poland).
Died: 15 February 1940 in Jerusalem (then under British Mandate).

After completing his secondary education in Breslau, Otto Toeplitz entered the university there to study mathematics. After graduating, he continued with his studies of algebraic geometry at the University of Breslau, being awarded his doctorate in 1905.

In 1906 Toeplitz went to Göttingen and studied under Hilbert who was working on the theory of integral equations at the time. Toeplitz began to rework the classical theories of forms on spaces of finite dimension to spaces of infinite dimension. He wrote five papers directly related to Hilbert's spectral theory. Also during this period he published a paper on summation processes and discovered the basic ideas of what are now called the 'Toeplitz operators'. When Toeplitz arrived in Göttingen, Hellinger was a doctoral student there. The two quickly became friends and they would collaborate closely for many years.

It was not until 1913 that Toeplitz was offered a teaching post as extraordinary professor at the University of Kiel. He was promoted to ordinary professor at Kiel in 1920. A joint project with Hellinger to write a major encyclopaedia article on integral equations, which they worked on for many years, was completed during this time and appeared in print in 1927. In 1928 Toeplitz accepted an offer of a chair at the University of Bonn.

Toeplitz was dismissed from his chair by the Nazis in 1935. He did not leave Germany but rather remained and worked hard to support the Jewish community who were finding life increasingly difficult. He emigrated to Palestine (as it then was) in 1939 and helped in the building up of Jerusalem University. He had great plans for modernizing the university but unfortunately he became very ill and died a year after his arrival.

Toeplitz worked on infinite linear and quadratic forms. In the 1930's he developed a general theory of infinite dimensional spaces and criticised Banach's work as being too abstract. In a joint paper with Köthe in 1934, Toeplitz introduced, in the context of linear sequence spaces, some important new concepts and theorems. Köthe has described how they would find the same idea cropping up in different contexts.

Toeplitz was also very interested in the history of mathematics. For example he wrote *The Calculus: A Genetic Approach*, an excellent book on the history of the calculus. It was originally published posthumously in German in 1949 edited by Köthe. A historical topic which interested him deeply was the relation between Greek mathematics and Greek philosophy. He was a frequent visitor to the Frankfurt Mathematics Seminar in the 1920s and 30s, where his friend Hellinger worked from 1914, and there the history of mathematics played a large role. Toeplitz believed that only a mathematician of stature is qualified to be a historian of mathematics.

It was not only the history of mathematics that interested him outside his research area. He wrote a popular book on mathematics in collaboration with Rademacher. This work, *The Enjoyment of Mathematics*, has been reprinted many times over the years. Toeplitz was also greatly interested in school mathematics and devoted much time to it.

Vandermonde, Alexandre Théophile

Born: 28 February 1735 in Paris, France.
Died: 1 January 1796 in Paris, France.

Alexandre-Théophile Vandermonde graduated *licencié* in 1757. His first love was music and his instrument was the violin. He pursued a music career and he only turned to mathematics when he was 35 years old. It was Fontaine des Bertins whose enthusiasm for mathematics rubbed off on Vandermonde. He was elected to the Académie des Sciences in 1771 after his first paper was read to the Academy. This and three further papers which he presented between 1771 and 1772 represent his total mathematical output.

In 1777 Vandermonde published the results of experiments he had carried out with Bezout and Lavoisier on the effects of a very severe frost which had occurred in 1776. Ten years later, with Monge and Bertholet, he published two papers on manufacturing steel. He was so close to Monge that he was often called 'la femme de Monge'. In 1778 he put forward the idea that musicians should ignore all theory and rely solely on their trained ears when judging music.

Positions which Vandermonde held include director of the Conservatoire des Arts et Métiers in 1782 and chief of the Bureau de l'Habillement des Armées in 1792. In the same year of 1792 he sat with Lagrange on a committee of the Académie des Sciences which had to examine a newly invented musical instrument. He was involved with designing a course in political economy for the École Normale, after it was founded in October 1794.

The name of Vandermonde is best known today for the Vandermonde determinant, yet it does not appear in his four mathematical papers. It is rather strange, therefore, that this determinant should be named after him. Lebesgue conjectured that it resulted from someone misreading Vandermonde's notation, and therefore believing that this determinant was in his work.

The first of Vandermonde's papers presented a formula for the sum of the m-th powers of the roots of an equation. It also presented a formula for the sum of the symmetric functions of the powers of such roots. Kronecker claimed in 1888 that the study of modern algebra began with this paper. Cauchy stated that Vandermonde had priority over Lagrange for ideas which eventually led to the study of group theory.

In his second paper Vandermonde considered the problem of the knight's tour on the chess board. He considered the intertwining of curves generated by the moving knight, an early example of topological ideas.

In his third paper Vandermonde studied combinatorial ideas, and in the final paper he studied the theory of determinants. Muir claims Vandermonde was 'the only one fit to be viewed as the founder of the theory of determinants'.

The reason for Muir's claim is that, although determinants had been studied earlier by Leibniz, all earlier work had simply used the determinant as a tool to solve linear equations. Vandermonde, however, thought of the determinant as a function and gave functional properties. He showed the effect of interchanging two rows and of interchanging two columns. From this he deduced that a determinant with two identical rows or two identical columns is zero.

1.1, 1.2 Routine verification of the axioms.

1.3 $\|x + y\|^2 = \langle x+y \mid x+y \rangle = \langle x \mid x \rangle + \langle y \mid x \rangle + \langle x \mid y \rangle + \langle y \mid y \rangle = \|x\|^2 + \|y\|^2 + 2\langle x \mid y \rangle$. If $x, y \in \mathbb{R}^2$ are linearly independent and if L is the line joining x to the origin then the line through y perpendicular to L meets L at the point λx where $\langle y - \lambda x \mid x \rangle = 0$. Clearly, $\lambda = \langle x \mid y \rangle / \|x\|^2$. If ϑ is the angle xOy then

$$\cos \vartheta = \frac{\|\lambda x\|}{\|y\|} = \frac{|\lambda| \, \|x\|}{\|y\|} = \frac{|\langle x \mid y \rangle|}{\|x\| \, \|y\|}$$

and so we have the cosine law $\|x + y\|^2 = \|x\|^2 + \|y\|^2 + 2\|x\| \, \|y\| \cos \vartheta$.

1.4 Expanding $\|x + y\|^2 - i \|ix + y\|^2$ in a complex inner product space we obtain

$$\|x\|^2 + \|y\|^2 + \langle x \mid y \rangle + \langle y \mid x \rangle - i[\, \|ix\|^2 + \|y\|^2 + \langle ix \mid y \rangle + \langle y \mid ix \rangle]$$
$$= \|x\|^2 + \|y\|^2 - i(\|x\|^2 + \|y\|^2) + 2\langle x \mid y \rangle.$$

1.5 $\|x + y\|^2 + \|x - y\|^2 = \|x\|^2 + \|y\|^2 + \langle x \mid y \rangle + \langle y \mid x \rangle + \|x\|^2 + \|-y\|^2 + \langle x \mid -y \rangle + \langle -y \mid x \rangle = 2\|x\|^2 + 2\|y\|^2$. In $\mathbb{R}^2$ this result says that the sum of the squares of the lengths of the diagonals of a parallelogram is the sum of the squares of the lengths of its sides.

1.6 If $\|x\| = \|y\|$ then

$$\langle x + y \mid x - y \rangle = \langle x \mid x \rangle + \langle y \mid x \rangle + \langle x \mid -y \rangle + \langle y \mid -y \rangle = \|x\|^2 - \|y\|^2 = 0.$$

In $\mathbb{R}^2$ this says that the diagonals of a rhombus are mutually perpendicular.

1.7 $\left| \int_a^b f(x) g(x) \, dx \right| \leqslant \left(\int_a^b |f(x)|^2 dx \right)^{1/2} \left(\int_a^b |g(x)|^2 dx \right)^{1/2}$.

1.8 $|\operatorname{tr} B^* A| \leqslant |\operatorname{tr} A^* A|^{1/2} |\operatorname{tr} B^* B|^{1/2}$.

$$\langle E_{pq} \mid E_{rs} \rangle = \operatorname{tr} E_{pq}^* E_{rs} = \operatorname{tr} E_{qp} E_{rs} = \begin{cases} 1 & \text{if } r = p, \ s = q; \\ 0 & \text{otherwise.} \end{cases}$$

1.9 If $y = \lambda x$ then clearly $|\langle x \mid y \rangle| = |\lambda| \, \|x\|^2 = \|x\| \, \|y\|$. Conversely, suppose that equality holds. We may assume that $y \neq 0$. Let $a = \langle x \mid y \rangle / \|y\|^2$. Let $z = x - ay$. Then, by the hypothesis, $\langle y \mid x \rangle = \langle y \mid ay \rangle + \langle y \mid z \rangle = \langle x \mid y \rangle + \langle y \mid z \rangle$. It follows that $\langle y \mid z \rangle = 0$. Since $|a| = \|x\| / \|y\|$ we then have

$$\|x\|^2 = \|ay + z\|^2 = |a|^2 \|y\|^2 + \|z\|^2$$

whence $\|z\|^2 = 0$ and consequently $z = 0$. Hence $x = ay$.

1.10 $a = \frac{1}{2}$. **1.11** This is a standard exercise in calculus. **1.12** Yes.

1.13 Let $x_1 = (1,1,1)$, $x_2 = (0,1,1)$, $x_3 = (0,0,1)$.
Take $y_1 = x_1/\|x_1\| = \frac{1}{\sqrt{3}}(1,1,1)$. Now

$$x_2 - \langle x_2\,|\,y_1\rangle y_1 = \tfrac{1}{3}(-2,1,1)$$

so take $y_2 = \frac{1}{\sqrt{6}}(-2,1,1)$. Next,

$$x_3 - \langle x_3\,|\,y_2\rangle y_2 - \langle x_3\,|\,y_1\rangle y_1 = \tfrac{1}{2}(0,1,-1)$$

so take $y_3 = \frac{1}{\sqrt{2}}(0,1,-1)$. An orthonormal basis is then $\{y_1,y_2,y_3\}$.
1.14 Let $x_1 = (1,1,0,1)$, $x_2 = (1,-2,0,0)$, $x_3 = (1,0,-1,2)$. Then $\{x_1,x_2,x_3\}$ is linearly independent.
Take $y_1 = x_1/\|x_1\| = \frac{1}{\sqrt{3}}(1,1,0,1)$. Now

$$x_2 - \langle x_2\,|\,y_1\rangle y_1 = \tfrac{1}{3}(4,-5,0,1)$$

so take $y_2 = \frac{1}{\sqrt{42}}(4,-5,0,1)$. Next,

$$x_3 - \langle x_3\,|\,y_2\rangle y_2 - \langle x_3\,|\,y_1\rangle y_1 = \tfrac{1}{7}(-4,-2,-1,6)$$

so take $y_3 = \frac{1}{\sqrt{57}}(-4,-2,-1,6)$. An orthonormal basis is then $\{y_1,y_2,y_3\}$.
1.15 Let $x_1 = (1,i,0)$, $x_2 = (1,2,1-i)$. Recall the standard inner product of Example 1.2.
Take $y_1 = x_1/\|x_1\| = \frac{1}{\sqrt{2}}(1,i,0)$. Now

$$x_2 - \langle x_2\,|\,y_1\rangle y_1 = (\tfrac{1}{2}+i,1-\tfrac{1}{2}i,1-i)$$

so take $y_2 = \frac{\sqrt{2}}{3}(\tfrac{1}{2}+i,1-\tfrac{1}{2}i,1-i)$. An orthonormal basis is then $\{y_1,y_2\}$.
1.16 We have $f_1 : t \mapsto 1$ and $f_2 : t \mapsto t$. Then

$$\langle f_1\,|\,f_1\rangle = \int_0^1 dt = 1$$

so we can take $g_1 = f_1$ as the first vector in the Gram–Schmidt process. Now

$$\langle f_2\,|\,g_1\rangle = \int_0^1 t\,dt = \tfrac{1}{2}$$

and so $f_2 - \langle f_2\,|\,g_1\rangle g_1$ is the mapping $h : t \mapsto t - \tfrac{1}{2}$. Since

$$\langle h\,|\,h\rangle = \int_0^1 [h(t)]^2 dt = \int_0^1 (t^2 - t + \tfrac{1}{4})dt = \tfrac{1}{12},$$

we can take g_2 to be the mapping $t \mapsto 2\sqrt{3}(t - \tfrac{1}{2})$. Then $\{g_1, g_2\}$ is an orthonormal basis.
1.17 (1) Observe that

$$\langle X\,|\,X^2\rangle = \int_{-1}^1 X^3 = 0$$

and so $\{X, X^2\}$ is orthogonal. It therefore suffices to normalise both vectors. Now

$$\langle X\,|\,X\rangle = \int_{-1}^1 X^2 = \tfrac{2}{3} \quad\text{and}\quad \langle X^2\,|\,X^2\rangle = \int_{-1}^1 X^4 = \tfrac{2}{5}$$

and so an orthonormal basis is $\{\frac{\sqrt{3}}{\sqrt{2}}X, \frac{\sqrt{5}}{\sqrt{2}}X^2\}$.
(2) Let $x_1 = 1$, $x_2 = 2X - 1$, $x_3 = 12X^2$. Then we have $\langle x_1\,|\,x_1\rangle = 2$ so we take $y_1 = x_1/\|x_1\| = \frac{1}{\sqrt{2}}$. Next,

$$\langle x_2\,|\,y_1\rangle = \frac{1}{\sqrt{2}}\int_{-1}^1 (2X-1) = -\sqrt{2}$$

and so $x_2 - \langle x_2 | y_1 \rangle y_1 = 2X - 1 + 1 = 2X$. So we take $y_2 = 2X/\|2X\| = \frac{\sqrt{3}}{\sqrt{2}}X$. Finally, $\langle x_3 | y_2 \rangle = 0$ and $\langle x_3 | y_1 \rangle = 4\sqrt{2}$ and so

$$x_3 - \langle x_3 | y_2 \rangle y_2 - \langle x_3 | y_1 \rangle y_1 = 4(3X^2 - 1)$$

and so we take $y_3 = \frac{\sqrt{5}}{2\sqrt{2}}(3X^2 - 1)$.

1.18 Let $\{e_1, e_2\}$ be the standard orthonormal basis of $\mathbb{R}^2$. Let x and y have polar coordinates r_1, ϑ_1 and r_2, ϑ_2 respectively. Then

$$\cos(\vartheta_2 - \vartheta_1) = \frac{\langle x | y \rangle}{r_1 r_2}$$

and similarly

$$\cos \vartheta_1 = \frac{\langle x | e_1 \rangle}{r_1}, \quad \sin \vartheta_1 = \frac{\langle x | e_2 \rangle}{r_1}.$$

Parseval's identity, namely

$$\langle x | y \rangle = \langle x | e_1 \rangle \langle e_1 | y \rangle + \langle x | e_2 \rangle \langle e_2 | y \rangle,$$

is then equivalent to

$$r_1 r_2 \cos(\vartheta_2 - \vartheta_1) = r_1 r_2 \cos \vartheta_1 \cos \vartheta_2 + r_1 r_2 \sin \vartheta_1 \sin \vartheta_2$$

from which we may remove the factor $r_1 r_2$.

1.19 Bessel's inequality becomes

$$\sum_{k=-n}^{n} \left| \int_{-\pi}^{\pi} f(x) e^{ikx} dx \right|^2 \leqslant \int_{-\pi}^{\pi} |f(x)|^2 dx.$$

The rest of the question is a standard exercise in calculus.

1.20 That B is an orthonormal basis is clear from the fact that B is linearly independent and orthonormal. If $(2, -3, 1) = \alpha(\frac{1}{\sqrt{5}}, 0, \frac{2}{\sqrt{5}}) + \beta(-\frac{2}{\sqrt{5}}, 0, \frac{1}{\sqrt{5}}) + \gamma(0, 1, 0)$ then the Fourier coefficients α, β, γ are given by

$$\alpha = \langle (2, -3, 1) | (\tfrac{1}{\sqrt{5}}, 0, \tfrac{2}{\sqrt{5}}) \rangle = \tfrac{3}{\sqrt{5}};$$

$$\beta = \langle (2, -3, 1) | (-\tfrac{2}{\sqrt{5}}, 0, \tfrac{1}{\sqrt{5}}) \rangle = -\tfrac{3}{\sqrt{5}};$$

$$\gamma = \langle (2, -3, 1) | (0, 1, 0) \rangle = -3.$$

1.21 If $M = [m_{ij}]_{n \times n}$ then for all i, j we have $f(e_j) = m_{ij} e_i$. Consequently

$$\langle f(e_j) | e_i \rangle = \langle m_{ij} e_i | e_i \rangle = m_{ij} \langle e_i | e_i \rangle = m_{ij}.$$

1.22 Suppose first that $a_1, \ldots, a_r$ are linearly dependent. Then some a_j is a linear combination of the a_k with $k \neq j$. Consequently, for every i, $\langle a_i | a_j \rangle$ is a linear combination of the $\langle a_i | a_k \rangle$ with $k \neq j$. Thus the j-th column of G_A is a linear combination of the other columns and therefore $\det G_A = 0$.

Suppose now that $a_1, \ldots, a_r$ are linearly independent. Then they form a basis for the subspace that they generate. By Gram–Schmidt, let $\{b_1, \ldots, b_r\}$ be an orthonormal basis for this subspace. Let M' be the transition matrix from the ordered basis $(a_i)_r$ to the ordered basis $(b_i)_r$. Then we have

$$\langle a_i | a_j \rangle = \left\langle \sum_{k=1}^{r} m_{ik} b_k \,\middle|\, \sum_{t=1}^{r} m_{jt} b_t \right\rangle = m_{i1} m_{j1} + m_{i2} m_{j2} + \cdots + m_{ir} m_{jr} = [MM']_{ij}.$$

Hence $G_A = MM'$ and so $\det G_A = (\det M)^2$. Since transition matrices are invertible, we have $\det M \neq 0$ whence $\det G_A \neq 0$.

1.23 (1) $\Rightarrow$ (2) : If f is an isomorphism then $\|f(x)\|^2 = \langle f(x)\,|\,f(x)\rangle = \langle x\,|\,x\rangle = \|x\|^2$.
(2) $\Rightarrow$ (1) : If (2) holds then

$$\|x + y\|^2 = \|f(x + y)\|^2 = \langle f(x + y)\,|\,f(x + y)\rangle$$
$$= \|x\|^2 + \langle f(x)\,|\,f(y)\rangle + \langle f(y)\,|\,f(x)\rangle + \|y\|^2.$$

Since

$$\|x + y\|^2 = \|x\|^2 + \langle x\,|\,y\rangle + \langle y\,|\,x\rangle + \|y\|^2$$

it follows that

$$\langle f(x)\,|\,f(y)\rangle + \langle f(y)\,|\,f(x)\rangle = \langle x\,|\,y\rangle + \langle y\,|\,x\rangle,$$

whence the result if $F = \mathbb{R}$. Doing the same with $\|ix + y\|^2$ we see similarly that

$$\langle f(x)\,|\,f(y)\rangle - \langle f(y)\,|\,f(x)\rangle = \langle x\,|\,y\rangle - \langle y\,|\,x\rangle.$$

It follows that $\langle f(x)\,|\,f(y)\rangle = \langle x\,|\,y\rangle$.
(2) $\Rightarrow$ (3) : If (2) holds then $\|f(x) - f(y)\| = \|f(x - y)\| = \|x - y\|$.
(3) $\Rightarrow$ (2) : If (3) holds then $\|f(x) - f(y)\| = \|x - y\|$ and so, on taking $y = 0$, we have $\|f(x)\| = \|x\|$.
1.24 We have $[\varphi(af + bg)](x) = x(af + bg)(x) = x[af(x) + bg(x)] = xaf(x) + xbg(x) = [\varphi(af)](x) + [\varphi(bg)](x)$ and so $\varphi(af + bg) = \varphi(af) + \varphi(bg)$. Thus φ is linear. It preserves inner products since

$$\langle \varphi(f)\,|\,\varphi(g)\rangle = \int_0^1 [\varphi(f)](x)[\varphi(g)](x)dx = \int_0^1 x^2 f(x)g(x)dx = \langle f\,|\,g\rangle.$$

However, it is not an isomorphism since it fails to be surjective. To see this, consider the constant function $g : x \mapsto 1$ in V. There is no $f \in V$ such that $xf(x) = 1$ for every $x \in [0, 1]$.
2.1 Let $U_1 = \{(0, a, b)\ ;\ a, b \in \mathbb{R}\}$ and $U_2 = \{(a, 0, b)\ ;\ a, b \in \mathbb{R}\}$. Then $V \cap U_1 = \{0\} = V \cap U_2$. Since

$$(a, b, c) = (a, a, 0) + (0, b - a, c) = (b, b, 0) + (a - b, 0, c)$$

we then have $\mathbb{R}^3 = V \oplus U_1 = V \oplus U_2$.
2.2 We have
  Ker $t_1 = \{(0, 0, 0)\}$ and so, by the dimension theorem, Im $t_1 = \mathbb{R}^3$.
  Ker $t_2 = \{(a, a, a)\ ;\ a \in \mathbb{R}\}$ and Im $t_2 = \{(a, b, 0)\ ;\ a, b \in \mathbb{R}\}$.
  Ker $t_3 = \{(0, 0, 0)\}$, Im $t_3 = \mathbb{R}^3$.
  Ker $t_4 = \{(0, 0, a)\ ;\ a \in \mathbb{R}\}$ and Im $t_4 = \{(a, b, b)\ ;\ a, b \in \mathbb{R}\}$.
For $i = 1, 2, 3, 4$ we see that Im $t_i \cap$ Ker $t_i = \{(0, 0, 0)\}$. It follows by the dimension theorem that $\dim(\text{Im } t_i + \text{Ker } t_i) = \dim \mathbb{R}^3$ and therefore Im $t_i + $ Ker $t_i = \mathbb{R}^3$. Hence $\mathbb{R}^3 = \text{Im } t_i \oplus \text{Ker } t_i$.
2.3 Let $\dim V = n$ and let $\{a_1, \ldots, a_r\}$ be a basis of A. Extend this to a basis

$$\{a_1, \ldots, a_r, b_1, \ldots, b_{n-r}\}$$

of V. Then the subspace B that is spanned by $\{b_1, \ldots, b_{n-r}\}$ is such that $A \oplus B = V$.
2.4 Every line L of non-zero gradient that passes through the origin is a subspace of $\mathbb{R}^2$ such that $X \oplus L = \mathbb{R}^2$.
2.5 The cartesian product $A \times B$ is a vector space with $\dim A \times B = \dim A + \dim B = \dim V$. The mapping $f : A \times B \to V$ given by $f(a, b) = a - b$ is linear with $\text{Im} f = A + B$ and Ker $f = \{(x, x)\ ;\ x \in A \cap B\} \simeq A \cap B$. By the dimension theorem we have $\dim A \times B = \dim \text{Im} f + \dim \text{Ker } f$. Thus $\dim V = \dim(A + B) + \dim(A \cap B)$. Since $A + B = V$, it follows that $\dim(A \cap B) = 0$ whence $A \cap B = \{0_V\}$. Hence $V = A \oplus B$.

2.6 Suppose that $V = \operatorname{Im} f \oplus \operatorname{Ker} f$. Clearly, $\operatorname{Im} f^2 \subseteq \operatorname{Im} f$. If now $x \in \operatorname{Im} f$ then $x = f(a)$ for some $a \in V$, and $a = p + q$ where $p \in \operatorname{Im} f$ and $q \in \operatorname{Ker} f$. Then $x = f(p + q) = f(p) + 0_V = f(p) \in \operatorname{Im} f^2$. Hence $\operatorname{Im} f \subseteq \operatorname{Im} f^2$ and we have equality.

Conversely, suppose that $\operatorname{Im} f = \operatorname{Im} f^2$. Then for every $x \in V$ we have $f(x) = f^2(y)$ for some $y \in V$. This gives $x - f(y) \in \operatorname{Ker} f$ and so $x = f(y) + [x - f(y)] \in \operatorname{Im} f + \operatorname{Ker} f$. Hence $V = \operatorname{Im} f + \operatorname{Ker} f$. Observe now from the dimension theorem that

$$\dim \operatorname{Ker} f = \dim V - \dim \operatorname{Im} f = \dim V - \dim \operatorname{Im} f^2 = \dim \operatorname{Ker} f^2.$$

Since $\operatorname{Ker} f \subseteq \operatorname{Ker} f^2$ we deduce that $\operatorname{Ker} f = \operatorname{Ker} f^2$. Suppose now that $x \in \operatorname{Im} f \cap \operatorname{Ker} f$. Then $x = f(y)$ and $f(x) = 0_V$. Consequently $f^2(y) = 0_V$ so $y \in \operatorname{Ker} f^2 = \operatorname{Ker} f$ whence $0_V = f(y) = x$. Hence $\operatorname{Im} f \cap \operatorname{Ker} f = \{0_V\}$ and therefore $V = \operatorname{Im} f \oplus \operatorname{Ker} f$.

2.7 It is clear that we have the chains

$$V \supseteq \operatorname{Im} f \supseteq \operatorname{Im} f^2 \supseteq \cdots \supseteq \operatorname{Im} f^n \supseteq \operatorname{Im} f^{m+1} \supseteq \cdots ;$$

$$\{0_V\} \subseteq \operatorname{Ker} f \subseteq \operatorname{Ker} f^2 \subseteq \cdots \subseteq \operatorname{Ker} f^n \subseteq \operatorname{Ker} f^{n+1} \subseteq \cdots .$$

Now we cannot have an infinite sumber of strict inclusions in the first chain since $X \subset Y$ implies that $\dim X < \dim Y$, and the dimension of V is finite. It follows that there is a smallest positive integer p such that $\operatorname{Im} f^p = \operatorname{Im} f^{p+1} = \cdots$. Since $\dim V = \dim \operatorname{Im} f^p + \dim \operatorname{Ker} f^p$ a corresponding result holds for the kernel chain. Writing $g = f^p$ we see that $\operatorname{Im} g = \operatorname{Im} g^2$. It then follows by the previous exercise that $V = \operatorname{Im} f^p \oplus \operatorname{Ker} f^p$.

2.8 (1) is routine.

(2) If $x \in V_1$ then $x = x_1(b_1 + b_4) + x_2(b_2 + b_3)$; so V_1 is spanned by $\{b_1 + b_4, b_2 + b_3\}$. Also, if $x_1(b_1 + b_4) + x_2(b_2 + b_3) = 0_V$ then since B is a basis of V we have $x_1 = x_2 = 0$. Thus $B_1 = \{b_1 + b_4, b_2 + b_3\}$ is a basis of V_1. Similarly, $B_2 = \{b_1 - b_4, b_2 - b_3\}$ is a basis of V_2.

(3) It is clear from the definitions that $V_1 \cap V_2 = \{0_V\}$. Consequently, the sum $V_1 + V_2$ is direct. Since V_1, V_2 are each of dimension 2 and V is of dimension 4, it follows that $V = V_1 \oplus V_2$.

(4) As for the transition matrix from the basis B to the basis $B_1 \cup B_2$, we observe that

$$b_1 = \tfrac{1}{2}(b_1 + b_4) + 0(b_2 + b_3) + 0(b_2 - b_3) + \tfrac{1}{2}(b_1 - b_4)$$
$$b_2 = 0(b_1 + b_4) + \tfrac{1}{2}(b_2 + b_3) + \tfrac{1}{2}(b_2 - b_3) + 0(b_1 - b_4)$$
$$b_3 = 0(b_1 + b_4) + \tfrac{1}{2}(b_2 + b_3) - \tfrac{1}{2}(b_2 - b_3) + 0(b_1 - b_4)$$
$$b_4 = \tfrac{1}{2}(b_1 + b_4) + 0(b_2 + b_3) + 0(b_2 - b_3) - \tfrac{1}{2}(b_1 - b_4).$$

The matrix in question is therefore

$$P = \begin{bmatrix} \tfrac{1}{2} & 0 & 0 & \tfrac{1}{2} \\ 0 & \tfrac{1}{2} & \tfrac{1}{2} & 0 \\ 0 & \tfrac{1}{2} & -\tfrac{1}{2} & 0 \\ \tfrac{1}{2} & 0 & 0 & -\tfrac{1}{2} \end{bmatrix}.$$

(5) It is readily seen that $2P^2 = I_4$ and so

$$P^{-1} = 2P = \begin{bmatrix} 1 & 0 & 0 & 1 \\ 0 & 1 & 1 & 0 \\ 0 & 1 & -1 & 0 \\ 1 & 0 & 0 & -1 \end{bmatrix}.$$

Suppose now that M is centro-symmetric; i.e. is of the form

$$\begin{bmatrix} a & b & c & d \\ e & f & g & h \\ h & g & f & e \\ d & c & b & a \end{bmatrix}.$$

Let φ be the linear mapping represented by M relative to the ordered basis B. Then the matrix of φ relative to the ordered basis C is given by PMP^{-1}, which is readily seen to be of the form

$$K = \begin{bmatrix} \alpha & \beta & 0 & 0 \\ \gamma & \delta & 0 & 0 \\ 0 & 0 & \epsilon & \zeta \\ 0 & 0 & \eta & \vartheta \end{bmatrix}.$$

Thus if M is centro-symmetric then it is similar to a matrix of the form K.

2.9 For every $x \in V$ we have

$$x = (\mathrm{id}_V + f)(\tfrac{1}{2}x) + (\mathrm{id}_V - f)(\tfrac{1}{2}x)$$

and so $V = \mathrm{Im}\,(\mathrm{id}_V + f) + \mathrm{Im}\,(\mathrm{id}_V - f)$. If now $x \in \mathrm{Im}\,(\mathrm{id}_V + f) \cap \mathrm{Im}\,(\mathrm{id}_V - f)$ then for some $y, z \in V$ we have $x = y + f(y) = z - f(z)$. Since by hypothesis $f^2 = \mathrm{id}_V$, it follows that

$$f(x) = f(y) + f^2(y) = f(y) + y = x;$$
$$f(x) = f(z) - f^2(z) = f(z) - z = -x.$$

Hence $x = 0_V$ and therefore $V = \mathrm{Im}\,(\mathrm{id}_V + f) \oplus \mathrm{Im}\,(\mathrm{id}_V - f)$.

If $A^2 = I_n$ let f be a linear mapping that is represented by A relative to some ordered basis of V. Then $f^2 = \mathrm{id}_V$. Let $\{a_1, \ldots, a_p\}$ be a basis of $\mathrm{Im}\,(\mathrm{id}_V + f)$ and let $\{a_{p+1}, \ldots, a_n\}$ be a basis of $\mathrm{Im}\,(\mathrm{id}_V - f)$. Then $\{a_1, \ldots, a_n\}$ is a basis of V. Now since $a_1 = b + f(b)$ for some $b \in V$ we have $f(a_1) = f(b) + f^2(b) = f(b) + b = a_1$, and similarly for $a_2, \ldots, a_p$. Likewise, $a_{p+1} = c - f(c)$ for some $c \in V$ so $f(a_{p+1}) = f(c) - f^2(c) = f(c) - c = -a_{p+1}$, and similarly for $a_{p+2}, \ldots, a_n$. Hence the matrix of f relative to the ordered basis $\{a_1, \ldots, a_n\}$ is

$$\Delta_p = \begin{bmatrix} I_p & 0 \\ 0 & -I_{n-p} \end{bmatrix},$$

and A is then similar to this matrix. Conversely, if A is similar to a matrix of the form Δ_p then there is an invertible matrix Q such that $Q^{-1}AQ = \Delta_p$. Consequently,

$$A^2 = (Q\Delta_p Q^{-1})^2 = Q\Delta_p^2 Q^{-1} = QI_n Q^{-1} = I_n.$$

2.10 Extend A to the basis $\{(1,0,1), (-1,1,2), (0,1,0)\}$ of $\mathbb{R}^3$. Then, relative to this basis we have $(1,2,1) = (1,0,1) + 2(0,1,0)$ and therefore $p_A(1,2,1) = (1,0,1)$.

2.11 If f is the projection on A parallel to B then for $x = a + b$ with $a \in A$ and $b \in B$ we have $f(x) = a$. Consequently $(\mathrm{id}_V - f)(x) = x - f(x) = x - a = b$ and so $\mathrm{id}_V - f$ is the projection on B parallel to A.

2.12 Clearly, $p_1 + p_2$ is a projection (=idempotent) if and only if, denoting composites by juxtaposition, $p_1 p_2 + p_2 p_1 = 0$. It is therefore clear that $(2) \Rightarrow (1)$.

$(1) \Rightarrow (2)$: Suppose that $p_1 + p_2$ is a projection. Multiplying each side of $p_1 p_2 + p_2 p_1 = 0$ on the left by p_1 we obtain $p_1 p_2 + p_1 p_2 p_1 = 0$; and multiplying each side on the right by p_1 we obtain $p_1 p_2 p_1 + p_2 p_1 = 0$. It follows that $p_1 p_2 = p_2 p_1$. But $p_1 p_2 + p_2 p_1 = 0$. Hence $p_1 p_2 = p_2 p_1 = 0$.

Suppose now that $p_1 + p_2$ is a projection. If $x \in \mathrm{Im}\,(p_1 + p_2)$ then there exists y such that $x = p_1(y) + p_2(y)$ whence $x \in \mathrm{Im}\,p_1 + \mathrm{Im}\,p_2$. Thus $\mathrm{Im}\,(p_1 + p_2) \subseteq \mathrm{Im}\,p_1 + \mathrm{Im}\,p_2$. The reverse inclusion holds since if $x = p_1(y) + p_2(z)$ then applying p_1 to this and using the fact that $p_1 p_2 = 0$ we obtain $p_1(x) = p_1(y)$, and similarly $p_2(x) = p_2(z)$. Hence $x = p_1(x) + p_2(x) = (p_1 + p_2)(x) \in \mathrm{Im}\,(p_1 + p_2)$. It suffices now to show that the sum $\mathrm{Im}\,p_1 + \mathrm{Im}\,p_2$ is direct. For this purpose, let $x \in \mathrm{Im}\,p_1 \cap \mathrm{Im}\,p_2$. Then $x = p_1(y) = p_2(z)$ for some y, z. Then $x = p_1(x) = p_1 p_2(z) = 0_V$.

If $x \in \mathrm{Ker}\,(p_1 + p_2)$ then $p_1(x) + p_2(x) = 0_V$. Applying p_1 and p_2 and using the fact that $p_1 p_2 = 0 = p_2 p_1$, we obtain $p_1(x) = 0_V = p_2(x)$. Hence $\mathrm{Ker}\,(p_1 + p_2) \subseteq \mathrm{Ker}\,p_1 \cap \mathrm{Ker}\,p_2$.

Conversely, if $x \in \operatorname{Ker} p_1 \cap \operatorname{Ker} p_2$ then $p_1(x) = 0_V = p_2(x)$ and so $(p_1 + p_2)(x) = 0_V$, whence we have the reverse inclusion.

2.13 (1) $\Rightarrow$ (2) : Suppose that $\operatorname{Im} p_1 = \operatorname{Im} p_2$. Then for every $x \in V$ there exists $y \in V$ such that $p_1(x) = p_2(y)$. Applying p_2 to this, we obtain $p_2 p_1(x) = p_2(y) = p_1(x)$, whence $p_2 p_1 = p_1$. Similarly, $p_1 p_2 = p_2$.

(2) $\Rightarrow$ (1) : If (2) holds then for every $x \in V$ we have $p_1(x) = p_2[p_1(x)] \in \operatorname{Im} p_2$ whence $\operatorname{Im} p_1 \subseteq \operatorname{Im} p_2$. Similarly, we have the reverse inclusion.

2.14 Since $\operatorname{Im} p_1 = \cdots = \operatorname{Im} p_k$ we have, by the previous exercise, $p_i p_j = p_j$ for all i, j. Now if
$$p = \sum_{i=1}^{k} \lambda_i p_i \text{ then}$$

$$
\begin{aligned}
p^2 &= \lambda_1^2 p_1^2 + \cdots + \lambda_k^2 p_k^2 + \lambda_1 \lambda_2 p_1 p_2 + \lambda_2 \lambda_1 p_2 p_1 + \cdots + \lambda_k \lambda_{k-1} p_k p_{k-1} \\
&= \Big(\sum_{i=1}^{k} \lambda_i \Big) \lambda_1 p_1 + \cdots + \Big(\sum_{i=1}^{k} \lambda_i \Big) \lambda_k p_k \\
&= p
\end{aligned}
$$

and so p is also a projection. To show that $\operatorname{Im} p = \operatorname{Im} p_1$ it suffices to prove that $p p_1 = p_1$ and $p_1 p = p$. As for the first of these, we have

$$p p_1 = (\lambda_1 p_1 + \cdots + \lambda_k p_k) p_1 = \lambda_1 p_1 + \cdots + \lambda_k p_1 = p_1,$$

and the second is obtained similarly.

2.15 $(\operatorname{id}_V + f)(\operatorname{id}_V - \frac{1}{2}f) = \operatorname{id}_V - \frac{1}{2}f + f - \frac{1}{2}f^2$ and since f is idempotent the right hand side is id_V. Hence $\operatorname{id}_V + f$ is invertible with inverse $\operatorname{id}_V - \frac{1}{2}f$.

2.16 The set $\{(1, 0, 1), (1, 2, -2)\}$ is linearly independent so can be extended to a basis of $\mathbb{R}^3$, for example

$$B = \{(1, 0, 1), (1, 2, -2), (1, 0, 0)\}.$$

An application of the Gram–Schmidt process yields the orthonormal basis

$$\{\tfrac{1}{\sqrt{2}}(1, 0, 1), \tfrac{1}{\sqrt{34}}(3, 4, -3), \tfrac{1}{\sqrt{17}}(2, -3, -2)\}.$$

Recall from the orthonormalisation process that the subspace spanned by $\{(1, 0, 1), (1, 2, -2)\}$ coincides with the subspace spanned by $\{\tfrac{1}{\sqrt{2}}(1, 0, 1), \tfrac{1}{\sqrt{34}}(3, 4, -3)\}$; and if we denote this subspace by W_1 then $\mathbb{R}^3 = W_1 \oplus W_2$ where $W_2 = \operatorname{Span}\{(2, -3, -2)\}$. Then W_2 is the desired orthogonal complement.

3.1 $f(x, 0, z, 0) = (x + 2z, 0, z, 0) \in W$.

3.2 If $f(x) = ax + b$ then

$$[\varphi(f)](x) = x \int_0^1 (at + b)\, dt = x(\tfrac{1}{2}a + b)$$

so $\varphi(f) \in W$.

3.3 (1) If $\operatorname{Im} p$ is f-invariant then for every $x \in V$ there exists $y \in V$ such that $fp(x) = p(y)$. Then $pfp(x) = p^2(y) = p(y) = fp(x)$ and so $pfp = fp$. Conversely, if $pfp = fp$ then for every $x \in V$ we have $fp(x) = pfp(x) \in \operatorname{Im} p$ so that $\operatorname{Im} p$ is f-invariant.

(2) By the Corollary to Theorem 2.4, $\operatorname{Ker} p$ is f-invariant if and only if $\operatorname{Im}(\operatorname{id}_V - p)$ is f-invariant. From the above, this is the case if and only if

$$(\operatorname{id}_V - p)f(\operatorname{id}_V - p) = (\operatorname{id}_V - p)f.$$

Expanding each side, we see that this is equivalent to $pfp = pf$.

3.4 If $f \circ g = \mathrm{id}_V$ then f is surjective, hence bijective since V is of finite dimension. Then $g = f^{-1}$ and so $g \circ f = \mathrm{id}_V$.

Suppose that W is a subspace of V that is f-invariant, so that $f^{\to}(W) \subseteq W$. Since f is an isomorphism we have $\dim f^{\to}(W) = \dim W$ and so $f^{\to}(W) = W$. Hence $W = g[f^{\to}(W)] = g(W)$ and so W is g-invariant. For the converse, interchange f and g.

The result is false for infinite-dimensional spaces. For example, consider the real vector space $\mathbb{R}[X]$ of polynomials over $\mathbb{R}$. Let f be the differentiation mapping and let g be the integration mapping. Each is linear and we have $f \circ g = \mathrm{id}$ but $g \circ f \neq \mathrm{id}$.

3.5 (a) Since $V = \bigoplus_{i=1}^{n} V_i$, every $x \in V$ can be expressed uniquely in the form $x = \sum_{i=1}^{n} x_i$

where $x_i \in V_i$ for each i. Then $f(x) = \sum_{i=1}^{n} f(x_i)$ where $f(x_i) \in V_i$ since by hypothesis each V_i is

f-invariant. Hence $\mathrm{Im}\, f = \sum_{i=1}^{n} \mathrm{Im}\, f_i$. Also, since each V_i is f-invariant, for each i we have

$$\mathrm{Im}\, f_i \cap \sum_{j \neq i} \mathrm{Im}\, f_j \subseteq V_i \cap \sum_{j \neq i} V_j = \{0_V\}.$$

Hence $\mathrm{Im}\, f = \bigoplus_{i=1}^{n} \mathrm{Im}\, f_i$.

(b) If now $x \in \mathrm{Ker}\, f$ then $0_V = f(x) = \sum_{i=1}^{n} f(x_i) = \sum_{i=1}^{n} f_i(x_i) \in \sum_{i=1}^{n} \mathrm{Im}\, f_i$. Since the sum is

direct it follows that each $f_i(x_i) = 0_V$ whence $x_i \in \mathrm{Ker}\, f_i$. Consequently, $x \in \sum_{i=1}^{n} \mathrm{Ker}\, f_i$ and so

$\mathrm{Ker}\, f = \sum_{i=1}^{n} \mathrm{Ker}\, f_i$. But $\mathrm{Ker}\, f_i \subseteq V_i$ for each i, and so

$$\mathrm{Ker}\, f_i \cap \sum_{j \neq i} \mathrm{Ker}\, f_j \subseteq V_i \cap \sum_{j \neq i} V_j = \{0_V\}.$$

It follows that $\mathrm{Ker}\, f = \bigoplus_{i=1}^{n} \mathrm{Ker}\, f_i$.

3.6 $f(x, y, z) = (2x + y - z, -2x - y + 3z, z)$. Relative to the standard ordered basis, the matrix of f is

$$A = \begin{bmatrix} 2 & 1 & -1 \\ -2 & -1 & 3 \\ 0 & 0 & 1 \end{bmatrix}.$$

It is readily seen that $c_A(X) = X(X - 1)^2 = m_A(X)$. Thus

$$\mathbb{R}^3 = \mathrm{Ker}\, f \oplus \mathrm{Ker}\, (f - \mathrm{id})^2$$

with $\mathrm{Ker}\, f$ of dimension 1 and $\mathrm{Ker}\, (f - \mathrm{id})^2$ of dimension 2. Now

$$f(x, y, z) = (0, 0, 0) \iff 2x + y = 0 = z$$

and so a basis for $\mathrm{Ker}\, f$ is $\{(-1, 2, 0)\}$. Also,

$$(f - \mathrm{id})(x, y, z) = (x + y - z, -2x - 2y + 3z, 0)$$

and therefore

$$(f - \mathrm{id})^2(x, y, z) = (-x - y + 2z, 2x + 2y - 4z, 0).$$

So a basis for $\mathrm{Ker}\, (f - \mathrm{id})^2$ is $\{(1, -1, 0), (1, 1, 1)\}$.

An ordered basis with respect to which the matrix of f is in block diagonal form is then

$$B = \{(-1, 2, 0), (1, -1, 0), (1, 1, 1)\}.$$

The transition matrix from B to the standard basis is

$$P = \begin{bmatrix} -1 & 1 & 1 \\ 2 & -1 & 1 \\ 0 & 0 & 1 \end{bmatrix}$$

and the block diagonal matrix that represents f relative to B is

$$P^{-1}AP = \begin{bmatrix} 0 & 0 & 0 \\ 0 & 1 & 1 \\ 0 & 0 & 1 \end{bmatrix}.$$

3.7 $f(x, y, z) = (x + y + z, x + y + z, x + y + z)$. Relative to the standard ordered basis, the matrix of f is

$$M = \begin{bmatrix} 1 & 1 & 1 \\ 1 & 1 & 1 \\ 1 & 1 & 1 \end{bmatrix}.$$

It is readily seen that $c_M(X) = -X^2(X - 3)$ and that $m_M(X) = X(X - 3)$. The matrix

$$N = \begin{bmatrix} 3 & 0 & 0 \\ 0 & 0 & 0 \\ 0 & 0 & 0 \end{bmatrix}$$

also has these characteristic and minimum polynomials. So M and N reduce to the same block diagonal form and therefore are similar.

3.8 If $f^3 = f$ then f satisfies the equation $X(X + 1)(X - 1) = 0$. The minimum polynomial of f must therefore be a product of distinct linear factors. Hence f is diagonalisable.

3.9 $f(x, y, z) = (-2x - y + z, 2x + y - 3z, -z)$. Relative to the standard ordered basis of $\mathbb{R}^3$ the matrix of f is

$$A = \begin{bmatrix} -2 & -1 & 1 \\ 2 & 1 & -3 \\ 0 & 0 & -1 \end{bmatrix}.$$

It is readily seen that the eigenvalues are 0 and -1, the latter being of algebraic multiplicity 2. The minimum polynomial is $X(X + 1)^2$. Consequently, f is not diagonalisable.

3.10 (a) There are three distinct eigenvalues, namely $2, 3, 6$ so f is diagonalisable.

(b) The minimum polynomial is $(X - 1)(X - 2)^2$ so f is not diagonalisable.

(c) The minimum polynomial is $(X - 2)^3$ so f is not diagonalisable.

4.1 $f(x, y, z) = (-x - y - z, 0, x + y + z)$ so $f^2(x, y, z) = f(-x - y - z, 0, x + y + z) = (0, 0, 0)$.

4.2 The matrix that represents f relative to the given basis is

$$A = \begin{bmatrix} -5 & 1 & 4 \\ -8 & 1 & 7 \\ -5 & 1 & 4 \end{bmatrix}.$$

Since $A^3 = 0$ we have that $f^3 = 0$ so f is nilpotent.

4.3 If λ is an eigenvalue of f then λ^n is an eigenvalue of $f^n = 0$.

4.4 $f_A(X) = AX - XA$ gives

$$f_A^2(X) = A^2X - 2AXA + XA^2; \quad f_A^3(X) = A^3X - 3A^2XA + 3AXA^2 - XA^3,$$

and so on. If therefore $A^n = 0$ we have $f_A^{n+1} = 0$.

4.5 $f(x, y, z) = (2x + y - z, -2x - y + 3z, z)$. A solution is provided as in Exercise 3.6, but here we proceed as in Examples 4.4 and 4.5. As before, we have

$$\mathbb{R}^3 = \operatorname{Ker} f \oplus \operatorname{Ker}(f - \operatorname{id})^2$$

with Ker f of dimension 1 and Ker $(f-\text{id})^2$ of dimension 2. Begin by choosing $v_1 \in \text{Ker} f$; as before we can take $v_1 = (-1, 2, 0)$.

Next, we have

$$(f - \text{id})(x, y, z) = (x + y - z, -2x - 2y + 3z, 0)$$

so we choose a non-zero $v_2 \in \text{Ker}(f-\text{id})$, for example $v_2 = (1, -1, 0)$. Next we choose v_3 independent of v_2 and such that $(f-\text{id})(v_3) = \alpha v_2$. For example (taking $\alpha = 1$) we choose $v_3 = (1, 1, 1)$.

Then $B = \{v_1, v_2, v_3\}$ is an ordered basis with respect to which the matrix of f is upper triangular.

4.6 $f(x, y, z) = (2x - 2y, x - y, -x + 3y + z)$. The matrix of f relative to the standard basis of $\mathbb{R}^3$ is

$$A = \begin{bmatrix} 2 & -2 & 0 \\ 1 & -1 & 0 \\ -1 & 3 & 1 \end{bmatrix}.$$

It is readily seen that $c_A(X) = X(X-1)^2 = m_A(X)$. Thus

$$\mathbb{R}^3 = \text{Ker} f \oplus \text{Ker}(f-\text{id})^2$$

with Ker f of dimension 1 and Ker $(f-\text{id})^2$ of dimension 2. Begin by choosing $v_1 \in \text{Ker} f$, say $v_1 = (1, 1, -2)$.

Now

$$(f - \text{id})(x, y, z) = (x - 2y, x - 2y, -x + 3y)$$

so we choose a non-zero $v_2 \in \text{Ker}(f-\text{id})$, for example $v_2 = (0, 0, 1)$. Next we choose v_3 independent of v_2 and such that $(f-\text{id})(v_3) = \alpha v_2$. For example (taking $\alpha = 1$) we choose $v_3 = (2, 1, 0)$.

An ordered basis relative to which the matrix of f is upper triangular is then

$$B = \{(1, 1, -2), (0, 0, 1), (2, 1, 0)\}.$$

4.7 In Exercise 4.5, relative to the ordered basis B the matrix of f along with its Jordan decomposition is

$$\begin{bmatrix} 0 & 0 & 0 \\ 0 & 1 & 1 \\ 0 & 0 & 1 \end{bmatrix} = \begin{bmatrix} 0 & 0 & 0 \\ 0 & 1 & 0 \\ 0 & 0 & 1 \end{bmatrix} + \begin{bmatrix} 0 & 0 & 0 \\ 0 & 0 & 1 \\ 0 & 0 & 0 \end{bmatrix} = D + N.$$

The transition matrix P from B to the standard basis is

$$P = \begin{bmatrix} -1 & 1 & 1 \\ 2 & -1 & 1 \\ 0 & 0 & 1 \end{bmatrix}; \qquad P^{-1} = \begin{bmatrix} 1 & 1 & -2 \\ 2 & 1 & -3 \\ 0 & 0 & 1 \end{bmatrix}.$$

Simple computations give

$$PDP^{-1} = \begin{bmatrix} 2 & 1 & -2 \\ -2 & -1 & 4 \\ 0 & 0 & 1 \end{bmatrix}, \qquad PNP^{-1} = \begin{bmatrix} 0 & 0 & 1 \\ 0 & 0 & -1 \\ 0 & 0 & 0 \end{bmatrix}.$$

Thus the diagonal and nilpotent parts of f are

$$d_f(x, y, z) = (2x + y - 2z, -2x - y + 4z, z);$$

$$n_f(x, y, z) = (z, -z, 0).$$

[As a check, note that $d_f + n_f = f$.]

Applying the same procedure to the linear mapping of Exercise 4.6, we obtain

$$d_f(x, y, z) = (2x - 2y, x - y, -2x + 4y + z);$$
$$n_f(x, y, z) = (0, 0, x - y).$$

4.8 The matrix of f relative to the basis $\{b_1, b_2, b_3\}$ is

$$\begin{bmatrix} -1 & 3 & 0 \\ 0 & 2 & 0 \\ 2 & 1 & -1 \end{bmatrix}.$$

It is readily seen that the minimum polynomial is $(X - 2)(X + 1)^2$. Hence

$$\mathbb{R}^3 = \text{Ker}\,(f - 2\text{id}) \oplus \text{Ker}\,(f + \text{id})^2$$

where $\text{Ker}\,(f - 2\text{id})$ is of dimension 1 and $\text{Ker}\,(f + \text{id})^2$ is of dimension 2.

First observe that $f(b_1 + b_2 + b_3) = 2(b_1 + b_2 + b_3)$ so we can choose $v_1 = b_1 + b_2 + b_3 \in \text{Ker}\,(f - 2\text{id})$.

Next, observe that

$$(f + \text{id})(b_1) = 2b_3;$$
$$(f + \text{id})(b_2) = 3b_1 + 3b_2 + b_3;$$
$$(f + \text{id})(b_3) = 0.$$

We require a non-zero $v_2 \in \text{Ker}\,(f + \text{id})$; so take $v_2 = b_3$. Next, we need to choose v_3 independent of v_2 and such that $(f + \text{id})(v_3) = \alpha v_2$. Taking $\alpha = 1$ we can choose $v_3 = b_1$.

An ordered basis with respect to which the matrix of f is upper triangular is then

$$B = \{b_1 + b_2 + b_3, b_3, b_1\}.$$

The transition matrix from B to the standard basis is

$$P = \begin{bmatrix} 1 & 0 & 1 \\ 1 & 0 & 0 \\ 1 & 1 & 0 \end{bmatrix}; \qquad P^{-1} = \begin{bmatrix} 0 & 1 & 0 \\ 0 & -1 & 1 \\ 1 & -1 & 0 \end{bmatrix}.$$

The matrix of f relative to B is

$$P^{-1}AP = \begin{bmatrix} 2 & 0 & 0 \\ 0 & -1 & 2 \\ 0 & 0 & -1 \end{bmatrix} = D + N.$$

Simple calculations give

$$PDP^{-1} = \begin{bmatrix} -1 & 3 & 0 \\ 0 & 2 & 0 \\ 0 & 3 & -1 \end{bmatrix}, \qquad PNP^{-1} = \begin{bmatrix} 0 & 0 & 0 \\ 0 & 0 & 0 \\ 2 & -2 & 0 \end{bmatrix}.$$

Thus the diagonal and nilpotent parts of f are (relative to the basis $\{b_1, b_2, b_3\}$)

$$d_f(b_1) = -b_1; \qquad\qquad n_f(b_1) = 2b_3;$$
$$d_f(b_2) = 3b_1 + 2b_2 + 3b_3; \qquad n_f(b_2) = -2b_3;$$
$$d_f(b_3) = -b_3. \qquad\qquad n_f(b_3) = 0.$$

4.9 For the given matrix

$$A = \begin{bmatrix} 0 & 0 & -1 \\ 1 & 0 & 1 \\ 0 & 1 & 1 \end{bmatrix}$$

the minimum polynomial is $(X+1)(X-1)^2$. If the linear mapping f is represented by A relative to the standard basis then a basis with respect to which the matrix of f is upper triangular is

$$B = \{(1,-2,1),(-1,0,1),(1,1,0)\},$$

the upper triangular form being

$$\begin{bmatrix} -1 & & \\ & 1 & 1 \\ & & 1 \end{bmatrix} = \begin{bmatrix} -1 & 0 & 0 \\ 0 & 1 & 0 \\ 0 & 0 & 1 \end{bmatrix} + \begin{bmatrix} 0 & 0 & 0 \\ 0 & 0 & 1 \\ 0 & 0 & 0 \end{bmatrix} = D + N.$$

The transition matrix P from B to the standard basis is

$$P = \begin{bmatrix} 1 & -1 & 1 \\ -2 & 0 & 1 \\ 1 & 1 & 0 \end{bmatrix}; \qquad P^{-1} = \tfrac{1}{4}\begin{bmatrix} 1 & -1 & 1 \\ -1 & 1 & 3 \\ 2 & 2 & 2 \end{bmatrix}.$$

Then

$$P^{-1}A^nP = (P^{-1}AP)^n = (D+N)^n = \begin{bmatrix} (-1)^n & 0 & 0 \\ 0 & 1 & n \\ 0 & 0 & 1 \end{bmatrix}$$

from which A^n can be obtained.

5.1 If $x \neq 0_V$ is such that $f^{p-1}(x) \neq 0_V$ then for every $k \leqslant p-1$ we have $f^k(x) \neq 0_V$. To show that $\{x, f(x), \ldots, f^{p-1}(x)\}$ is linearly independent, suppose that

$$\lambda_0 x + \lambda_1 f(x) + \cdots + \lambda_{p-1}f^{p-1}(x) = 0.$$

Applying f^{p-1} to this we obtain $\lambda_0 f^{p-1}(x) = 0$ whence $\lambda_0 = 0$. Thus we have

$$\lambda_1 f(x) + \cdots + \lambda_{p-1}f^{p-1}(x) = 0.$$

Applying f^{p-2} to this we obtain similarly $\lambda_1 = 0$. Continuing in this way we see that every $\lambda_i = 0$ and consequently the set is linearly independent.

5.2 If f is nilpotent of index $n = \dim V$ then $\{x, f(x), \ldots, f^{n-1}(x)\}$ is a basis of V. The matrix of f relative to this ordered basis is then that in the question. Conversely, if there is an ordered basis with respect to which the matrix of f is of the given form then to see that f is nilpotent of index n it suffices to observe that the matrix M in question is such that $M^n = 0$ and $M^{n-1} \neq 0$.

5.3 The eigenvalues are 2 of algebraic multiplicity 1, and 1 of algebraic multiplicity 3. The rank of the matrix $M - I_4$ is 3 so for any linear mapping f represented by the matrix we have $\dim \text{Ker}(f - \text{id}) = 4 - 3 = 1$. Thus there is a single Jordan block associated with the eigenvalue 1. The Jordan form is then

$$\begin{bmatrix} 1 & 1 & 0 & \\ 0 & 1 & 1 & \\ 0 & 0 & 1 & \\ & & & 2 \end{bmatrix}.$$

5.4 The eigenvalues are 2 of algebraic multiplicity 3, and 3 of algebraic multiplicity 2. The rank of the matrix $M - 2I_5$ is 3 so for any linear mapping f represented by the matrix we have $\dim \text{Ker}(f - 2\text{id}) = 5 - 3 = 2$. Thus there are two Jordan blocks associated with the eigenvalue

2. Similarly there is only one Jordan block associated with the eigenvalue 3. Hence the Jordan form is

$$\begin{bmatrix} 2 & 1 & & & \\ 0 & 2 & & & \\ & & 2 & & \\ & & & 3 & 1 \\ & & & 0 & 3 \end{bmatrix}.$$

5.5 A basis is $\{1, X, X^2, X^3\}$, relatve to which the matrix of D is

$$\begin{bmatrix} 0 & 1 & 0 & 0 \\ 0 & 0 & 2 & 0 \\ 0 & 0 & 0 & 3 \\ 0 & 0 & 0 & 0 \end{bmatrix}.$$

The characteristic and minimum polynomials are X^4 so the Jordan form is

$$\begin{bmatrix} 0 & 1 & 0 & 0 \\ 0 & 0 & 1 & 0 \\ 0 & 0 & 0 & 1 \\ 0 & 0 & 0 & 0 \end{bmatrix}.$$

5.6 We have

$$\varphi(x^2) = 2xy, \quad \varphi(xy) = \tfrac{1}{2}y^2, \quad \varphi(y^2) = 0, \quad \varphi(x) = y, \quad \varphi(y) = 0, \quad \varphi(1) = 0$$

and therefore

$$M = \begin{bmatrix} 0 & 0 & 0 & 0 & 0 & 0 \\ 2 & 0 & 0 & 0 & 0 & 0 \\ 0 & \tfrac{1}{2} & 0 & 0 & 0 & 0 \\ 0 & 0 & 0 & 0 & 0 & 0 \\ 0 & 0 & 0 & 1 & 0 & 0 \\ 0 & 0 & 0 & 0 & 0 & 0 \end{bmatrix}.$$

The characteristic polynomial of M is X^6 and the minimum polynomial is X^3.

The Jordan form of M therefore has the eigenvalue 0 six times down the diagonal with at least one Jordan block of size 3×3. The number of Jordan blocks is dim Ker φ. Clearly, a basis for Ker φ is $\{1, y, y^2\}$. Thus the Jordan form is

$$J = \begin{bmatrix} 0 & 1 & 0 & & & \\ 0 & 0 & 1 & & & \\ 0 & 0 & 0 & & & \\ & & & 0 & 1 & \\ & & & 0 & 0 & \\ & & & & & 0 \end{bmatrix}.$$

5.7 (1) Let A be the matrix of Exercise 5.4 and let $f_A : \text{Mat}_{5 \times 1} \mathbb{R} \to \text{Mat}_{5 \times 1} \mathbb{R}$ be the linear mapping whose matrix relative to the standard basis of $\text{Mat}_{5 \times 1} \mathbb{R}$ is A, so that $f_A(\mathbf{x}) = A\mathbf{x}$.

We now proceed as in Example 5.8. Begin by choosing $\mathbf{v}_1 \in \text{Im}\,(f - 2\text{id}) \cap \text{Ker}\,(f - 2\text{id})$, e.g. $\mathbf{v}_1 = [1\ 0\ 1\ 0\ 0]$. Next, choose $\mathbf{v}_2$ independent of $\mathbf{v}_1$ such that $(f - 2\text{id})(\mathbf{v}_2) = \mathbf{v}_1$, e.g. $\mathbf{v}_2 = [0\ 1\ 0\ 1\ 0]$. Next, choose $\mathbf{v}_3 \in \text{Ker}(f_A - 2\text{id})$ such that $\{\mathbf{v}_1, \mathbf{v}_2, \mathbf{v}_3\}$ is linearly independent, e.g. $\mathbf{v}_3 = [2\ 1\ 0\ 0\ 1]$. Now repeat the process with $f - 3\text{id}$ to obtain, for example, $\mathbf{v}_4 = [-1\ 0\ 0\ 1\ 0]$ and $\mathbf{v}_5 = [2\ 0\ 0\ 0\ 1]$. A Jordan basis is then

$$\{[1\ 0\ 1\ 0\ 0], [0\ 1\ 0\ 1\ 0], [2\ 1\ 0\ 0\ 1], [-1\ 0\ 0\ 1\ 0], [2\ 0\ 0\ 0\ 1]\}.$$

(2) A Jordan basis is $\{f_1, f_2, f_3, f_4\}$ where

$$Df_1 = 0, \quad Df_2 = f_1, \quad Df_3 = f_2, \quad Df_4 = f_3.$$

Choose $f_1 = 1$. Then $f_2 = X, f_3 = \frac{1}{2}X^2, f_4 = \frac{1}{6}X^3$. So a Jordan basis is $\{6, 6X, 3X^2, X^3\}$.

5.8 (1) To find a Jordan basis, first choose $v_1 \in \operatorname{Im} \varphi \cap \operatorname{Ker} \varphi$, e.g. $v_1(x, y) = \frac{1}{2}y^2$. Next choose v_2 independent of v_1 and such that $\varphi(v_2) = v_1$, e.g. $v_2(x, y) = xy$. Next choose v_3 independent of v_1, v_2 with $\varphi(v_3) = v_2$, e.g. $v_3(x, y) = \frac{1}{2}x^2$.

Now choose $v_4 \in \operatorname{Im} \varphi \cap \operatorname{Ker} \varphi$ independent of v_1, v_2, v_3, e.g. $v_4(x, y) = y$. Next choose v_5 independent of v_1, v_2, v_3, v_4 with $\varphi(v_5) = v_4$, e.g. $v_5(x, y) = x$.

Finally, choose $v_6 \in \operatorname{Ker} \varphi$ independent of $v_1, \ldots, v_5$, e.g. $v_6(x, y) = 1$.

Then a Jordan basis is $B = \{\frac{1}{2}y^2, xy, \frac{1}{2}x^2, y, x, 1\}$.

(2) The transition matrix from B to the given basis is

$$P = \begin{bmatrix} 0 & 0 & \frac{1}{2} & 0 & 0 & 0 \\ 0 & 1 & 0 & 0 & 0 & 0 \\ \frac{1}{2} & 0 & 0 & 0 & 0 & 0 \\ 0 & 0 & 0 & 0 & 1 & 0 \\ 0 & 0 & 0 & 1 & 0 & 0 \\ 0 & 0 & 0 & 0 & 0 & 1 \end{bmatrix}$$

and is such that $P^{-1}AP = J$.

5.9 The given matrix A has characteristic polynomial $X^3(X-1)^2$, so the eigenvalues are 0 and 1. The transition matrix from the given basis of $\mathbb{R}^5$ to the standard basis is

$$P = \begin{bmatrix} 1 & 1 & 1 & 1 & 1 \\ 0 & 1 & 1 & 1 & 1 \\ 0 & 0 & 1 & 1 & 1 \\ 0 & 0 & 0 & 1 & 1 \\ 0 & 0 & 0 & 0 & 1 \end{bmatrix}; \quad P^{-1} = \begin{bmatrix} 1 & -1 & 0 & 0 & 0 \\ 0 & 1 & -1 & 0 & 0 \\ 0 & 0 & 1 & -1 & 0 \\ 0 & 0 & 0 & 1 & -1 \\ 0 & 0 & 0 & 0 & 1 \end{bmatrix}.$$

The matrix of f relative to the standard basis is then

$$M = PAP^{-1} = \begin{bmatrix} 1 & 3 & -1 & -1 & -2 \\ 0 & -4 & 1 & 1 & 1 \\ 0 & -5 & 2 & 1 & 1 \\ 0 & -6 & 1 & 2 & 2 \\ 0 & -5 & 1 & 1 & 1 \end{bmatrix}.$$

Now $M - 1I_5$ has (column) rank 3 so the number of Jordan blocks associated with the eigenvalue 1 is $5 - 3 = 2$. Also, the rank of the matrix $M - 0I_5$ is 4 so the number of Jordan blocks associated with the eigenvalue 0 is $5 - 4 = 1$. Hence the Jordan form is

$$\begin{bmatrix} 1 & & & & \\ & 1 & & & \\ & & 0 & 1 & \\ & & & 0 & 1 \\ & & & & 0 \end{bmatrix}.$$

To compute a Jordan basis we begin by choosing two independent vectors v_1, v_2 in the kernel of $M - I_5$, say $v_1 = [1\ 0\ 0\ 0\ 0]$ and $v_2 = [0\ 0\ 1\ -1\ 0]$. Next, we choose a vector v_3 in both the image and the kernel of $M = M - 0I_5$, say $v_3 = [1\ 0\ 0\ -1\ 1]$; then a vector v_4 independent of v_3 and such that $Mv_4 = v_3$, say $v_4 = [0\ -1\ -1\ -5\ 2]$; then a vector v_5 independent of v_3, v_4 such that $Mv_5 = v_4$, say $v_5 = [1\ -3\ -3\ -20\ 10]$. The required Jordan basis is then $\{v_1, \ldots, v_5\}$.

5.10 The characteristic polynomial of A is $(X-2)^4$ and the minimum polynomial is $(X-2)^2$. Since $A - 2I_4$ has rank 1 the number of Jordan blocks associated with the (single) eigenvalue 2 is $4-1=3$. Hence the Jordan form is

$$J = \begin{bmatrix} 2 & 1 & & \\ & 2 & & \\ & & 2 & \\ & & & 2 \end{bmatrix}.$$

Now choose $v_1 \in \text{Im}\,(A-2I_4) \cap \text{Ker}\,(A-2I_4)$, say $v_1 = [-2\ -2\ -2\ 2]$, and then v_2 independent of v_1 and such that $(A - 2I_4)v_2 = v_1$, say $v_2 = [1\ 0\ 0\ 0]$. Finally, choose v_3 and v_4 such that $\{v_1, \ldots, v_4\}$ forms a basis, say $v_3 = [0\ 1\ 0\ 1]$ and $v_4 = [0\ 0\ 1\ 0]$. Then we form the matrix

$$P = \begin{bmatrix} -2 & 1 & 0 & 0 \\ -2 & 0 & 1 & 0 \\ -2 & 0 & 0 & 1 \\ 2 & 0 & 1 & 0 \end{bmatrix}.$$

To solve the system $x' = Ax$ we first solve the system $y' = Jy$, namely

$$y_1' = 2y_1 + y_2; \quad y_2' = 2y_2; \quad y_3' = 2y_3; \quad y_4' = 2y_4.$$

The solution is clearly

$$y_4 = c_4 e^{2t}; \quad y_3 = c_3 e^{2t}; \quad y_2 = c_2 e^{2t}; \quad y_1 = c_2 t e^{2t} + c_1 e^{2t}.$$

Since now $x = Py$ we deduce that

$$x_1 = -2c_2 t e^{2t} - 2c_1 e^{2t} + c_2 e^{2t};$$

$$x_2 = -2c_2 t e^{2t} - 2c_1 e^{2t} + c_3 e^{2t};$$

$$x_3 = -2c_2 t e^{2t} - 2c_1 e^{2t} + c_4 e^{2t};$$

$$x_4 = 2c_2 t e^{2t} + 2c_1 e^{2t} + c_3 e^{2t}.$$

5.11 The system is $x' = Ax$ where

$$A = \begin{bmatrix} 1 & 3 & -2 \\ 0 & 7 & -4 \\ 0 & 9 & -5 \end{bmatrix}.$$

Now A has Jordan form

$$J = \begin{bmatrix} 1 & 1 & 0 \\ 0 & 1 & 0 \\ 0 & 0 & 1 \end{bmatrix}$$

and an invertible matrix P such that $P^{-1}AP = J$ is

$$P = \begin{bmatrix} 3 & 0 & 1 \\ 6 & 1 & 0 \\ 9 & 0 & 0 \end{bmatrix}.$$

First solve $y' = Jy$ to obtain

$$y_1 = ae^t + bte^t; \quad y_2 = be^t; \quad y_3 = ce^t.$$

Then $x' = Ax$ has the general solution

$$x = Py = a[3\ 6\ 9]e^t + b([3\ 6\ 9]te^t + [0\ 1\ 0]e^t) + c[1\ 0\ 0]e^t.$$

5.12 The characteristic polynomial of A is $(X+2)(X-2)^2$ which is also the minimum polynomial. The Jordan form is

$$J = \begin{bmatrix} 2 & 1 & 0 \\ 0 & 2 & 0 \\ 0 & 0 & -2 \end{bmatrix}.$$

A Jordan basis is

$$\{[1\ 2\ 4], [0\ 1\ 4], [1\ -2\ 4]\}$$

and so an invertible matrix P such that $P^{-1}AP = J$ is

$$P = \begin{bmatrix} 1 & 0 & 1 \\ 2 & 1 & -2 \\ 4 & 4 & 4 \end{bmatrix}.$$

Now solve the system $\mathbf{y}' = J\mathbf{y}$ to get

$$y_1 = c_1 e^{2t} + c_2 t e^{2t}; \quad y_2 = c_2 e^{2t}; \quad y_3 = c_3 e^{-2t}.$$

Then $\mathbf{x} = P\mathbf{y}$ gives

$$x = x_1 = c_1 e^{2t} + c_2 t e^{2t} + c_3 e^{-2t}.$$

Now apply the initial conditions to get $x = (4t - 1)e^{2t} + e^{-2t}$.

6.1 If $Z \subseteq \mathrm{Ker}\, f$ then $x + Z = y + Z$ gives $x - y \in Z \subseteq \mathrm{Ker}\, f$ whence $f(x) = f(y)$. Hence the assignment $x + Z \mapsto f(x)$ defines a mapping $\vartheta_f : V/Z \to W$. This mapping is linear since $\vartheta_f[(x + Z) + (y + Z)] = \vartheta_f(x + y + Z) = f(x + y) = f(x) + f(y) = \vartheta_f(x + Z) + \vartheta_f(y + Z)$ and $\vartheta_f[\lambda(x + Z)] = \vartheta_f(\lambda x + Z) = f(\lambda x) = \lambda f(x) = \lambda \vartheta_f(x + Z)$.

Conversely, if the prescription $x + Z \mapsto f(x)$ defines a (linear) mapping $\vartheta_f : V/Z \to W$ then $x \in Z$ gives $x + Z = 0 + Z$ whence $f(x) = f(0) = 0$ and therefore $x \in \mathrm{Ker}\, f$. Hence $Z \subseteq \mathrm{Ker}\, f$.

6.2 Taking $Z = \mathrm{Ker}\, f$ in the previous exercise we see that the assignment $\vartheta : x + \mathrm{Ker}\, f \mapsto f(x)$ is a linear mapping from $V/\mathrm{Ker}\, f$ to W with $\mathrm{Im}\, \vartheta = \mathrm{Im}\, f$. Observe now that ϑ is injective since $\vartheta(x + \mathrm{Ker}\, f) = \vartheta(y + \mathrm{Ker}\, f)$ if and only if $f(x) = f(y)$, i.e. if and only if $x - y \in \mathrm{Ker}\, f$, which is equivalent to $x + \mathrm{Ker}\, f = y + \mathrm{Ker}\, f$. Hence ϑ is an isomorphism from $V/\mathrm{Ker}\, f$ onto $\mathrm{Im}\, f$.

6.3 $\mathrm{Ker}\, f = \{(0, y, z) \, ; \, y, z \in \mathbb{R}\}$ (the y, z-plane). Now

$$f^+[(x, y, z) + \mathrm{Ker}\, f] = f(x, y, z) + \mathrm{Ker}\, f = (x, x, x) + \mathrm{Ker}\, f = (x, y, z) + \mathrm{Ker}\, f$$

since $(x, x, x) - (x, y, z) \in \mathrm{Ker}\, f$, and so f^+ is the identity on $\mathbb{R}^3/\mathrm{Ker}\, f$.

6.4 $f(a, b, c) = (0, a, b)$ so $f(1, -1, 3) = (0, 1, -1)$ and $f^2(1, -1, 3) = f(0, 1, -1) = (0, 0, 1)$. Since then $f^3(1, -1, 3) = (0, 0, 0)$ it follows that

$$(a + bf + cf^2)(1, -1, 3) = (a, -a + b, 3a - b + c)$$

and so $Z_{(1,-1,3)} = \{(a, -a + b, 3a - b + c) \, ; \, a, b, c \in \mathbb{R}\}$.

6.5 In the previous exercise the f-annihilator of $(1, -1, 3)$ is X^3.

6.6 $f(x, y, z) = (x + z, y, -x - z)$. Then $f(1, 1, 1) = (2, 1, -2)$, $f^2(1, 1, 1) = (0, 1, 0)$, $f^3(1, 1, 1) = (0, 1, 0) = f^2(1, 1, 1)$. Thus the f-annihilator of $(1, 1, 1)$ is $X^3 - X^2$.

6.7 We have $c_A = m_A = (X - a)^3$. As in Example 6.5, we have $3 = n_1 + \cdots + n_k$ with $n_1 = 3$. Thus $k = 1$ and the rational form is

$$C_{(X-a)^3} = \begin{bmatrix} 0 & 0 & a^3 \\ 1 & 0 & -3a^2 \\ 0 & 1 & 3a \end{bmatrix}.$$

6.8 $c_A = (X + 1)^2(X - 5)$ and $m_A = (X + 1)(X - 5)$. The rational canonical form is then $\mathrm{diag}\{-1, -1, 5\}$ and coincides with the Jordan form.

6.9 $c_A = (X - 1)(X - 2)^2$ and $m_A = (X - 1)(X - 2)$. The rational canonical form is then $\text{diag}\{1, 2, 2\}$ and coincides with the Jordan form.

6.10 It is given that $c_f = (X - 1)^3(X - 2)^4$ and $m_f = (X - 1)^2(X - 2)^3$. Then $V = V_1 \oplus V_2$ where $\dim V_1 = 3$ and $\dim V_2 = 4$. The induced mapping f_1 on V_1 has minimum polynomial $(X - 1)^2$. By Corollary 2 of Theorem 6.6 we have $3 = n_1 + \cdots + n_k$ with $n_1 = 2$. So $k = 2$ and $n_2 = 1$. Likewise the induced mapping f_2 on V_2 has minimum polynomial $(X - 2)^3$. Then $4 = m_1 + \cdots + m_k$ with $m_1 = 3$. So $k = 2$ and $m_2 = 1$. Hence the rational canonical form is

$$C_{(X-1)^2} \oplus C_{X-1} \oplus C_{(X-2)^3} \oplus C_{X-2} = \begin{bmatrix} 0 & -1 & & & & & \\ 1 & 2 & & & & & \\ & & 1 & & & & \\ & & & 0 & 0 & 8 & \\ & & & 1 & 0 & -12 & \\ & & & 0 & 1 & 6 & \\ & & & & & & 2 \end{bmatrix}.$$

6.11 V has an f-cyclic vector α if and only if $Z_\alpha = V$, which is the case if and only if $\dim Z_\alpha = \dim V$, i.e. if and only if $\deg m_\alpha = \deg c_f$. But $\deg m_\alpha \leqslant \deg m_f \leqslant \deg c_f$ so if this holds then $\deg m_f = \deg c_f$, whence $m_f = c_f$.

For the converse it suffices to consider the case where $m_f = p^t$. As in the proof of Theorem 6.6, there is a non-zero vector x such that $m_x = p^t$. Then if $m_f = c_f$ we have

$$\dim Z_x = \deg m_x = \deg c_f = \dim V$$

whence $Z_x = V$ and so x is an f-cyclic vector for V.

6.12 Follow the hint.

6.13 (1) $c_A = m_A = X^3$. Then as usual $3 = (n_1 + \cdots + n_k)3$ with $n_1 = 1$, so $k = 1$. The rational form is then

$$C_{X^3} = \begin{bmatrix} 0 & 0 & 0 \\ 1 & 0 & 0 \\ 0 & 1 & 0 \end{bmatrix}.$$

The classical form is the same as the Jordan form, namely

$$\begin{bmatrix} 0 & 1 & 0 \\ 0 & 0 & 1 \\ 0 & 0 & 0 \end{bmatrix}.$$

(2) $c_A = X^2(X - 5)$ and $m_A = X(X - 5)$. For $p_1 = X$ we have $2 = (n_1 + \cdots + n_k)1$ with $n_1 = 1$, so $k = 2$ and $n_2 = 1$. For $p_2 = X - 5$ we have $1 = (n_1 + \cdots + n_k)1$ with $n_1 = 1$, so $k = 1$. The rational form is then

$$C_X \oplus C_X \oplus C_{X-5} = \begin{bmatrix} 0 & 0 & 0 \\ 0 & 0 & 0 \\ 0 & 0 & 5 \end{bmatrix}$$

which is the same as the classical form and the Jordan form.

(3) $c_A = X^2(X + 2)(X - 2)$ and $m_A = X(X + 2)(X - 2)$. For $p_1 = X$ we have $2 = (n_1 + \cdots + n_k)1$ with $n_1 = 1$. Hence $k = 2$ and $n_2 = 1$. Then for $p_2 = X - 2$ we have $1 = (n_1 + \cdots + n_k)1$ with $n_1 = 1$. Hence $k = 1$. Similarly for $p_3 = X + 2$. The rational form is then

$$C_X \oplus C_X \oplus C_{X-2} \oplus C_{X+2} = \text{diag}\{0, 0, 2, -2\}$$

which is the same as the classical form and the Jordan form.

6.14 $V = V_1 \oplus V_2$ where $\dim V_1 = 6$ and $\dim V_2 = 4$. The restriction f_1 of f to V_1 has minimum polynomial $(X^2 - X + 1)^2$. Then, as usual, $6 = (n_1 + \cdots + n_k)2$ with $n_1 = 2$ and so $k = 2$ and $n_2 = 1$. The restriction f_2 of f to V_2 has minimum polynomial $(X + 1)^2$. Then, as usual, $4 = (n_1 + \cdots + n_k)1$ with $n_1 = 2$ and so either $k = 2$ and $n_2 = 2$, or $k = 3$ and $n_2 = n_3 = 1$. The rational form is then one of

$$C_{(X^2-X+1)^2} \oplus C_{X^2-X+1} \oplus C_{(X+1)^2} \oplus C_{(X+1)^2};$$

$$C_{(X^2-X+1)^2} \oplus C_{X^2-X+1} \oplus C_{(X+1)^2} \oplus C_{X+1} \oplus C_{X+1}.$$

The classical form can be obtained from this as in Example 6.6.

6.15 In the complex case we have $X^2 - X + 1 = (X - \alpha)(X - \beta)$ where $\alpha = \frac{1}{2}(1 + i\sqrt{3})$ and $\beta = \frac{1}{2}(1 - i\sqrt{3})$. The rational form is the same as before, with $C_{(X^2-X+1)^2} \oplus C_{X^2-X+1}$ replaced by $C_{(X-\alpha)^2} \oplus C_{X-\alpha} \oplus C_{(X-\beta)^2} \oplus C_{X-\beta}$.

In this case all the eigenvalues belong to the ground field $\mathbb{C}$ and so the classical form is the same as the Jordan form, which is either

$$\begin{bmatrix} \alpha & 1 & & & & & & & & \\ 0 & \alpha & & & & & & & & \\ & & \alpha & & & & & & & \\ & & & \beta & 1 & & & & & \\ & & & 0 & \beta & & & & & \\ & & & & & \beta & & & & \\ & & & & & & -1 & 1 & & \\ & & & & & & 0 & -1 & & \\ & & & & & & & & -1 & 1 \\ & & & & & & & & 0 & -1 \end{bmatrix}$$

or the same matrix with the 1 in the $(9, 10)$ position replaced by 0.

7.1 Consider $E^d_{i,j} : \text{Mat}_{m \times n} F \to F$ defined by $E^d_{i,j}(A) = a_{ij}$ for each $A = [a_{ij}]_{m \times n}$. Clearly, $E^d_{i,j}$ is linear and

$$E^d_{i,j}(E_{p,q}) = \begin{cases} 1 & \text{if } p = i, \ q = j; \\ 0 & \text{otherwise.} \end{cases}$$

Hence $\{E^d_{i,j} \ ; \ i = 1, \ldots, m \text{ and } j = 1, \ldots, n\}$ is the dual basis.

7.2 That φ_B is a linear form is routine.

Suppose now that $\varphi \in V^d$. Then relative to the dual basis of the previous exercise, we have

$$\varphi = \sum_{j=1}^{n} \sum_{i=1}^{m} \varphi(E_{i,j}) E^d_{i,j}.$$

Let $B = [b_{ij}] \in \text{Mat}_{m \times n}$ be defined by $b_{ij} = \varphi(E_{i,j})$. Then

$$\varphi(A) = \sum_{j=1}^{n} \sum_{i=1}^{m} \varphi(E_{i,j}) E^d_{i,j}(A) = \sum_{j=1}^{n} \sum_{i=1}^{m} b_{ij} a_{ij} = \sum_{j=1}^{n} [B^t A]_{jj} = \text{tr } B^t A.$$

7.3 The transition matrix from the given basis to the standard basis is

$$P = \begin{bmatrix} 1 & -1 & 0 \\ 0 & 1 & 1 \\ -1 & 0 & 1 \end{bmatrix}, \qquad P^{-1} = \begin{bmatrix} \frac{1}{2} & \frac{1}{2} & -\frac{1}{2} \\ -\frac{1}{2} & \frac{1}{2} & -\frac{1}{2} \\ \frac{1}{2} & \frac{1}{2} & \frac{1}{2} \end{bmatrix}.$$

Hence the dual basis is

$$\{[\tfrac{1}{2}, \tfrac{1}{2}, -\tfrac{1}{2}], \ [-\tfrac{1}{2}, \tfrac{1}{2}, -\tfrac{1}{2}], \ [\tfrac{1}{2}, \tfrac{1}{2}, \tfrac{1}{2}]\}.$$

7.4 $\{[2, -1, 1, 0], [7, -3, 1, -1], [-10, 5, -2, 1], [-8, 3, -3, 1]\}$.

7.5 The transition matrix from the basis B to the basis A is

$$P = \begin{bmatrix} 1 & 2 \\ 3 & 4 \end{bmatrix}, \qquad P^{-1} = \begin{bmatrix} -2 & 1 \\ \frac{3}{2} & -\frac{1}{2} \end{bmatrix}.$$

Consequently, $B^d = \{-2\varphi_1 + \frac{3}{2}\varphi_2, \varphi_1 - \frac{1}{2}\varphi_2\}$.

7.6 $\{(4, 5, -2, 11), (3, 4, -2, 6), (2, 3, -1, 4), (0, 0, 0, 1)\}$.

7.7 Proceed as in Example 7.9. If $p = a_0 + a_1 X + a_2 X^2$ then

$$(\lambda_1 \varphi_1 + \lambda_2 \varphi_2 + \lambda_3 \varphi_3)(p) = (\lambda_1 + \lambda_3)a_0 + (\tfrac{1}{2}\lambda_1 + \lambda_2)a_1 + (\tfrac{1}{3}\lambda_1 + 2\lambda_2)a_2.$$

Then $\lambda_1 \varphi_1 + \lambda_2 \varphi_2 + \lambda_3 \varphi_3 = 0$ if and only if $\lambda_1 = \lambda_2 = \lambda_3 = 0$, so that $\{\varphi_1, \varphi_2, \varphi_3\}$ is linearly independent and so is a basis of $(\mathbb{R}_2[X])^d$. Since $\mathbb{R}_2[X] \simeq \mathbb{R}^3$ under the correspondence

$$a_0 + a_1 X + a_2 X^2 \leftrightarrow (a_0, a_1, a_2)$$

we have that $\mathbb{R}_2[X]^d \simeq (\mathbb{R}^3)^d$. Now

$$\varphi_1(p) = a_0 + \tfrac{1}{2}a_1 + \tfrac{1}{3}a_2$$

$$\varphi_2(p) = a_1 + 2a_2$$

$$\varphi_3(p) = a_0$$

and so we can associate with $\{\varphi_1, \varphi_2, \varphi_3\}$ the basis

$$B = \{[1, \tfrac{1}{2}, \tfrac{1}{3}], [0, 1, 2], [1, 0, 0]\}$$

of $(\mathbb{R}^3)^d$. By considering the matrix

$$P = \begin{bmatrix} 1 & 0 & 1 \\ \frac{1}{2} & 1 & 0 \\ \frac{1}{3} & 2 & 0 \end{bmatrix}$$

whose inverse is

$$P^{-1} = \tfrac{3}{2} \begin{bmatrix} 0 & 2 & -1 \\ 0 & -\frac{1}{3} & \frac{1}{2} \\ \frac{2}{3} & -2 & 1 \end{bmatrix}$$

we see that

$$\{(0, 3, -\tfrac{3}{2}), (0, -\tfrac{1}{2}, \tfrac{3}{4}), (1, -3, \tfrac{3}{2})\}$$

is the basis of $\widehat{\mathbb{R}^3} = \mathbb{R}^3$ that is dual to B. Hence

$$\{3X - \tfrac{3}{2}X^2, -\tfrac{1}{2}X + \tfrac{3}{4}X^2, 1 - 3X + \tfrac{3}{2}X^2\}$$

is the basis of $\mathbb{R}^2[X]$ that is dual to the basis $\{\varphi_1, \varphi_2, \varphi_3\}$.

7.8 Properties (a) and (b) are immediate. If $\{\alpha_1, \ldots, \alpha_n\}$ is the basis dual to $\{\varphi_1, \ldots, \varphi_n\}$ then the solution set C consists of those linear combinations $t_1\alpha_1 + \cdots + t_n\alpha_n$ for which

$$\varphi_i(t_1\alpha_1 + \cdots + t_n\alpha_n) \geqslant 0.$$

But $\varphi_i = \alpha_i^d$ and so $\varphi_i(t_1\alpha_1 + \cdots + t_n\alpha_n) = t_i$.

7.9 $f : V \to V$ is surjective [injective] if and only if there exists a linear mapping $g : V \to V$ such that $f \circ g = \mathrm{id}_V$ [$g \circ f = \mathrm{id}_V$]. The result is therefore an immediate consequence of Theorem 7.5(1)(3).

7.10 That φ is linear follows from Theorem 7.5. That φ is injective follows from the fact that $f' = g'$ gives, relative to some fixed ordered basis of V, Mat $f' =$ Mat g' whence Mat $f =$ Mat g and therefore $f = g$. Since V is finite-dimensional, so is $\mathrm{Lin}\,(V, V)$. Hence φ is also surjective.

7.11 To find Y, we determine the dual of $\{[1, 0, 0], [1, 1, 0], [1, 1, 1]\}$. The transition matrix is

$$P = \begin{bmatrix} 1 & 1 & 1 \\ 0 & 1 & 1 \\ 0 & 0 & 1 \end{bmatrix}, \qquad P^{-1} = \begin{bmatrix} 1 & -1 & 0 \\ 0 & 1 & -1 \\ 0 & 0 & 1 \end{bmatrix}.$$

Thus $Y = \{(1, -1, 0), (0, 1, -1), (0, 0, 1)\}$.

The matrix of f with respect to the standard basis is

$$\begin{bmatrix} 2 & 1 & 0 \\ 1 & 1 & 1 \\ 0 & 0 & -1 \end{bmatrix}$$

and the transition matrices relative to X, Y are respectively

$$\begin{bmatrix} 1 & 0 & 0 \\ -1 & 1 & 0 \\ 0 & -1 & 1 \end{bmatrix}, \qquad \begin{bmatrix} 1 & 1 & 1 \\ 0 & 1 & 1 \\ 0 & 0 & 1 \end{bmatrix}.$$

By the change of basis theorem, the matrix of f relative to X, Y is then

$$\begin{bmatrix} 1 & 0 & 0 \\ 1 & 1 & 0 \\ 1 & 1 & 1 \end{bmatrix} \begin{bmatrix} 2 & 1 & 0 \\ 1 & 1 & 1 \\ 0 & 0 & -1 \end{bmatrix} \begin{bmatrix} 1 & 1 & 1 \\ 0 & 1 & 1 \\ 0 & 0 & 1 \end{bmatrix} = \begin{bmatrix} 2 & 3 & 3 \\ 3 & 5 & 6 \\ 3 & 5 & 5 \end{bmatrix}.$$

The required matrix is then the transpose of this.

7.12 For every linear form $\vartheta : \mathbb{R}_n[X] \to \mathbb{R}$ we have $D'(\vartheta) = \vartheta \circ D$. Thus $\vartheta \in \mathrm{Ker}\, D'$ if and only if $\vartheta \circ D = 0$ which is the case if and only if $\vartheta^{\to}(\mathbb{R}_{n-1}[X]) = 0$. Hence a basis for $\mathrm{Ker}\, D'$ is $\{D^n\}$.

7.13 Observe that $A \subseteq B$ implies $B^\circ \subseteq A^\circ$ so for the first equality we need only show that $\bigcap V_i^\circ \subseteq (\sum V_i)^\circ$. But if $f \in \bigcap V_i^\circ$ then f annihilates each V_i whence it annihilates $\sum V_i$. The second equality can be obtained from the first by applying the first equality to the family of dual spaces and identifying V with its bidual.

7.14 $f - g \in W^\circ$ if and only if $(\forall w \in W)\ f(w) = g(w)$, i.e. $f|_W = g|_W$. The linear mapping $V^d \to W^d$ described by $f \mapsto f|_W$ therefore has kernel W°. Now apply the first isomorphism theorem.

7.15 $S^{\circ\circ}$ is a subspace of V with $S \subseteq S^{\circ\circ}$. If now $W = \mathrm{Span}\, S$ then $S \subseteq W$ gives $S \subseteq S^{\circ\circ} \subseteq W^{\circ\circ} = W$. Hence $S^{\circ\circ} = W$.

7.16 Every linear form $f : \mathbb{R}^4 \to \mathbb{R}$ is given by

$$f(x_1, x_2, x_3, x_4) = a_1 x_1 + a_2 x_2 + a_3 x_3 + a_4 x_4$$

where $a_1 = f(1, 0, 0, 0)$, etc.. In order to have $f \in W^\circ$ we require $a_1 + a_2 = 0 = a_3 + a_4$, so that f is given by

$$f(x_1, x_2, x_3, x_4) = a_1 x_1 - a_1 x_2 + a_3 x_3 - a_3 x_4.$$

Now the dimension of W° is $4 - 2 = 2$ and so we can obtain a basis for W° by first taking $a_1 = 1, a_3 = 0$ then $a_1 = 0, a_3 = 1$. A basis is therefore $\{f_1, f_2\}$ where

$$f_1(x_1, x_2, x_3, x_4) = x_1 - x_2, \quad f_2(x_1, x_2, x_3, x_4) = x_3 - x_4.$$

7.17 $\pi'_A : A^d \to V^d$ is given by $\pi'_A(f) = f\pi_A$, and similarly for π'_B. For every $k \in V^d$ we have

$$k = k\,\mathrm{id}_V = k(\pi_A + \pi_B) = k|_A \pi_A + k|_B \pi_B = \pi'_A(k|_A) + \pi'_B(k|_B) \in \mathrm{Im}\, \pi'_A + \mathrm{Im}\, \pi'_B.$$

Thus $V^d = \operatorname{Im} \pi'_A + \operatorname{Im} \pi'_B$.

If now $k \in \operatorname{Im} \pi'_A \cap \operatorname{Im} \pi'_B$ then there exist f, g with $k = f\pi_A$ and $k = g\pi_B$. Consequently,

$$0 = f0 = f\pi_A\pi_B = k\pi_B = g\pi_B\pi_B = g\pi_B = k.$$

Hence $\operatorname{Im} \pi'_A \cap \operatorname{Im} \pi'_B = \{0\}$ and therefore $V^d = \operatorname{Im} \pi'_A \oplus \operatorname{Im} \pi'_B$.

7.18 Follow the hints.

7.19 That $f_{x,y}$ is linear is routine. As for (a), we have

$$f_{x,y}[f_{y,z}(t)] = f_{x,y}(\langle t \mid z\rangle y) = \langle\langle t \mid z\rangle y \mid y\rangle x = \langle t \mid z\rangle \langle y \mid y\rangle x = \|y\|^2 \langle t \mid z\rangle x = \|y\|^2 f_{x,z}(t).$$

As for (b), we have

$$\langle f_{x,y}(z) \mid t\rangle = \langle\langle z \mid y\rangle x \mid t\rangle = \langle z \mid y\rangle \langle x \mid t\rangle = \langle z \mid \overline{\langle x \mid t\rangle}y\rangle = \langle z \mid \langle t \mid x\rangle y\rangle = \langle z \mid f_{y,x}(t)\rangle.$$

7.20 (a) From the previous exercise we have that $f_{x,y}$ is normal if and only if it commutes with $f_{y,x}$, which is the case if and only if, for all $z \in V$,

$$\|y\|^2 \langle z \mid x\rangle x = \|x\|^2 \langle z \mid y\rangle y.$$

Taking $z = x$ in this we obtain

$$\|y\|^2 \|x\|^2 x = \|x\|^2 \langle x \mid y\rangle y$$

which gives $x = \lambda y$ where $\lambda = \langle x \mid y\rangle / \|y\|^2 \in \mathbb{C}$, so that the condition is necessary. Conversely, if $x = \lambda y$ for some $\lambda \in \mathbb{C}$ then

$$\|y\|^2 \langle z \mid x\rangle x = \|y\|^2 \langle z \mid y\rangle\overline{\lambda}\,\lambda y = |\lambda|^2 \|y\|^2 \langle z \mid y\rangle y = \|x\|^2 \langle z \mid y\rangle y$$

so that the condition is also sufficient.

(b) From the previous exercise we have that $f_{x,y}$ is self-adjoint if and only if, for all $z \in V$,

$$\langle z \mid y\rangle x = \langle z \mid x\rangle y.$$

Taking $z = y$ we obtain $x = \lambda y$ where $\lambda = \langle y \mid x\rangle / \|y\|^2$. This gives $\langle x \mid y\rangle = \lambda \|y\|^2 = \langle y \mid x\rangle$, whence we see that $\lambda \in \mathbb{R}$. The converse is clear.

7.21 It is readily seen that $\langle K(f) \mid g\rangle$ and $\langle f \mid K(g)\rangle$ are each equal to

$$\int_0^1 yf(y)\,dy \int_0^1 xg(x)\,dx.$$

7.22 We have

$$\begin{aligned}
\langle(a,b,c) \mid f^*(x,y,z)\rangle &= \langle f(a,b,c) \mid (x,y,z)\rangle \\
&= \langle(a+b,b,a+b+c) \mid (x,y,z)\rangle \\
&= x(a+b) + yb + z(a+b+c) \\
&= a(x+z) + b(x+y+z) + cz \\
&= \langle(a,b,c) \mid (x+z, x+y+z, z)\rangle.
\end{aligned}$$

Hence $f^*(x,y,z) = (x+z, x+y+z, z)$.

7.23 For the β given in the hint, we have $\langle \alpha_k \mid \beta\rangle = f(\alpha_k)$ for each basis element α_k.

$\beta = (1, -2, 4)$.

7.24 Follow the hint.

7.25 We have

$$\langle f_p(q) \mid r\rangle = \int_0^1 pq\bar{r} = \int_0^1 q\overline{\bar{p}r} = \langle q \mid f_{\bar{p}}(r)\rangle$$

and so $f_p^* = f_{\bar{p}}$.

For the rest of the question, follow the hint.

7.26 This is immediate from Theorem 7.11(1)(2)(4).

7.27 If W is f-invariant then

$$x \in W^{\perp} \Rightarrow (\forall w \in W) \; 0 = \langle f(w) \,|\, x \rangle = \langle w \,|\, f^*(x) \rangle \Rightarrow f^*(x) \in W^{\perp}$$

so $W^{\perp}$ is f^*-invariant.

If $W^{\perp}$ is f^*-invariant then

$$w \in W \Rightarrow (\forall y \in W^{\perp}) \; 0 = \langle w \,|\, f^*(y) \rangle = \langle f(w) \,|\, y \rangle \Rightarrow f(w) \in W^{\perp\perp} = W$$

so W is f-invariant.

7.28 $A^* = \overline{A^t} = \overline{A}^t$ and therefore $\det A^* = \det \overline{A} = \overline{\det A}$.

8.1 An orthonormal basis for W is $\{\frac{1}{\sqrt{5}}(0,1,2,0), \frac{1}{\sqrt{2}}(1,0,0,1)\}$. The Fourier coefficients of $(1,2,0,0)$ are $\frac{2}{\sqrt{5}}$ and $\frac{1}{\sqrt{2}}$ and so

$$p_W(1,2,0,0) = \tfrac{2}{5}(0,1,2,0) + \tfrac{1}{2}(1,0,0,1) = (\tfrac{1}{2}, \tfrac{2}{5}, \tfrac{4}{5}, \tfrac{1}{2}).$$

8.2 We have $V = W \oplus W^{\perp}$. Let $\{e_1, \ldots, e_k\}$ be an orthonormal basis for W and extend this to an orthonormal basis $\{e_1, \ldots, e_k, e_{k+1}, \ldots, e_n\}$ of V. Then $\{e_{k+1}, \ldots, e_n\}$ is an orthonormal basis of $W^{\perp}$. For every $x \in V$ we then have

$$x = \sum_{i=1}^{k} \langle x \,|\, e_i \rangle x + \sum_{i=k+1}^{n} \langle x \,|\, e_i \rangle x = p_W(x) + p_{W^{\perp}}(x)$$

whence $p_{W^{\perp}} = \mathrm{id}_V - p_W$.

8.3 Suppose that W is of dimension n. Then, for every $x \in V$,

$$\|p_W(x)\|^2 = \Big\langle \sum_{i=1}^{n} \langle x \,|\, e_i \rangle e_i \,\Big|\, \sum_{i=1}^{n} \langle x \,|\, e_i \rangle e_i \Big\rangle = \sum_{i=1}^{n} |\langle x \,|\, e_i \rangle|^2 \leqslant \|x\|^2.$$

8.4 The matrix representing a rotation through an angle ϑ is

$$R_\vartheta = \begin{bmatrix} \cos \vartheta & -\sin \vartheta \\ \sin \vartheta & \cos \vartheta \end{bmatrix}$$

and the matrix that represents a projection onto the x-axis is

$$A = \begin{bmatrix} 1 & 0 \\ 0 & 0 \end{bmatrix}.$$

The required matrix representing p_L is then $R_\vartheta A R_{-\vartheta}$.

8.5 See Example 4.5. The eigenspaces are $E_{-1} = \mathrm{Span}\{(1,-2,1)\}$ and $E_1 = \mathrm{Span}\{(-1,0,1)\}$. The ortho-projection on E_{-1} is given by

$$\langle (x,y,z) \,|\, \tfrac{1}{\sqrt{6}}(1,-2,1) \rangle \tfrac{1}{\sqrt{6}}(1,-2,1) = \tfrac{1}{6}(x + 2y - z)(1,-2,1)$$

and similarly that on E_1 is $\frac{1}{2}(-x + z)(-1,0,1)$.

8.6 $A^*A = AA^*$; take inverses, noting that $(A^*)^{-1} = (A^{-1})^*$.

8.7 Under the given conditions, we have

$$(A + iB)(A + iB)^* = A^2 + iBA - iAB + B^2$$

$$(A + iB)^*(A + iB) = A^2 - iBA + iAB + B^2.$$

Hence $A + iB$ is normal if and only if A, B commute.

8.8 Observe that $A^* = A$ and $B^* = -B$. Moreover, A and B do not commute.

8.9 $f_1^* = \frac{1}{2}(f^* + f^{**}) = \frac{1}{2}(f^* + f)$ so f_1 is self-adjoint. Also, $f_2^* = -\frac{1}{2i}(f^* - f^{**}) = -\frac{1}{2i}(f^* - f) = f_2$ so f_2 is also self-adjoint.

It is clear that $f = f_1 + if_2$. If also $f = g_1 + ig_2$ where g_1, g_2 are self-adjoint then we have $f_1 - g_1 = i(g_2 - f_2)$. Consequently, $f_1^\star - g_1^\star = -i(g_2 - f_2)^\star$, i.e. $f_1 - g_1 = -i(g_2 - f_2)$. It follows that $f_1 - g_1 = 0 = g_2 - f_2$.

$f^\star = f_1^\star - if_2^\star = f_1 - if_2$. Hence f is normal if and only if $f_1 + if_2$ commutes with $f_1 - if_2$. This is the case if and only if f_1 commutes with f_2.

8.10 Relative to the standard basis the matrix A of f is

$$A = \begin{bmatrix} 2 & i \\ 1 & 2 \end{bmatrix}; \quad A^\star = \begin{bmatrix} 2 & 1 \\ -i & 2 \end{bmatrix}.$$

Clearly, A is not self-adjoint. Since

$$AA^\star = \begin{bmatrix} 5 & 2+2i \\ 2-2i & 5 \end{bmatrix} = A^\star A$$

we see that A is normal.

8.11 This follows by a simple application of Theorem 7.11.

8.12 Either verify directly that $AA^\star = A^\star A$ or observe that $A = iB + B^\star$ where

$$B = \begin{bmatrix} 1 & i & -2i \\ 0 & 2 & 0 \\ i & 0 & 3 \end{bmatrix}$$

and use Exercise 8.11.

8.13 Suppose that A is normal and that $B = g(A)$. There is a unitary matrix P and a diagonal matrix D such that $P^{-1}AP = D$. Then

$$B = g(A) = Pg(D)P^{-1}.$$

Consequently, $\overline{B}^t B = Pg(\overline{D})g(D)P^{-1}$ and similarly $B\overline{B}^t = Pg(D)g(\overline{D})P^{-1}$. Since $g(D)$ and $g(\overline{D})$ are diagonal matrices it follows that $\overline{B}^t B = B\overline{B}^t$ and so B is normal.

8.14 If $A^\star = p(A)$ then clearly $AA^\star = A^\star A$, so that A is normal. Conversely, suppose that A is normal. Then there is a unitary matrix P and a diagonal matrix D such that $A = PDP^{-1}$. Let $\lambda_1, \ldots, \lambda_r$ be the distinct diagonal elements of D. Then there exist unique $a_1, \ldots, a_r$ such that

$$\overline{\lambda_1} = a_0 + a_1\lambda_1 + a_2\lambda_1^2 + \cdots + a_{r-1}\lambda_1^{r-1}$$
$$\overline{\lambda_2} = a_0 + a_1\lambda_2 + a_2\lambda_2^2 + \cdots + a_{r-1}\lambda_2^{r-1}$$
$$\vdots$$
$$\overline{\lambda_r} = a_0 + a_1\lambda_r + a_2\lambda_r^2 + \cdots + a_{r-1}\lambda_r^{r-1}.$$

[In fact, since $\lambda_1, \ldots, \lambda_r$ are distinct, the (Vandermonde) coefficient matrix is non-singular and so the system has a unique solution.] We then have

$$\overline{D} = a_0I + a_1D + a_2D^2 + \cdots + a_{r-1}D^{r-1}$$

and consequently

$$A^\star = P\overline{D}P^{-1} = a_0I + a_1A + \cdots + a_{r-1}A^{r-1}.$$

8.15 Consider matrices relative to the standard ordered basis. The matrices in $(1), (2), (4)$ are self-adjoint, hence normal. That in (3) is not normal.

8.16 Taking adjoints in $A^\star A = -A$ we obtain $-A^\star = A^\star A^{\star\star} = A^\star A$ whence $A^\star = A$ and so A is self-adjoint.

Let $\lambda_1, \ldots, \lambda_r$ be the distinct non-zero eigenvalues of A. From the above we have $A^2 = -A$, from which it follows that the distinct non-zero eigenvalues of $-A$, namely $-\lambda_1, \ldots, -\lambda_r$ are precisely the distinct non-zero eigenvalues of A^2, namely $\lambda_1^2, \ldots, \lambda_r^2$. Consequently, $\lambda_1, \ldots, \lambda_r$ are all negative. Let $\alpha_i = -\lambda_i > 0$ for each i, and suppose that

$$\alpha_1 < \alpha_2 < \cdots < \alpha_r.$$

Then this chain must coincide with the chain

$$\alpha_1^2 < \alpha_2^2 < \cdots < \alpha_r^2.$$

Consequently, $\alpha_i = \alpha_i^2$ for every i and so each $\alpha_i = 1$. Since by hypothesis $\alpha_i \neq 0$ we deduce that $\lambda_i = -\alpha_i = -1$.

8.17 Suppose that A is unitary, so that $A^\star = A^{-1}$. Then $A^\star A = I = AA^\star$ and so A is normal. Also (regarding A as a linear mapping) we have

$$\langle Ax \mid Ax \rangle = \langle x, A^\star Ax \rangle = \langle x, x \rangle$$

whence $\|Ax\| = \|x\|$. If now λ is any eigenvalue of A we have $\|Ax\| = |\lambda| \|x\|$. Thus we have $|\lambda| = 1$.

Conversely, suppose that A is normal and that every eigenvalue of A is of modulus 1. Let $f : V \to V$ be an ortho-diagonalisable mapping that is represented by A (cf. Theorem 8.8). If the spectral resolution of f is $f = \sum_{i=1}^{k} \lambda_i p_i$ then we know that the spectral resolution of $f^\star$ is $\sum_{i=1}^{k} \overline{\lambda_i} p_i$ and that

$$f \circ f^\star = \sum_{i=1}^{k} |\lambda_i|^2 p_i = f^\star \circ f.$$

If each eigenvalue of f is of modulus 1 it follows that

$$f \circ f^\star = f^\star \circ f = \sum_{i=1}^{k} p_i = \mathrm{id}_V.$$

Hence $f^\star = f^{-1}$ and consequently A is unitary.

8.18 The eigenvalues of A are $-1, 1, 8$. Corresponding general eigenvectors are

$$\begin{bmatrix} 0 \\ x \\ 0 \end{bmatrix}, \quad \begin{bmatrix} x \\ 0 \\ x \end{bmatrix}, \quad \begin{bmatrix} x \\ 0 \\ -x \end{bmatrix} \quad (x \neq 0).$$

An orthogonal matrix P such that $P^{-1}AP$ is diagonal is then

$$P = \begin{bmatrix} 0 & \frac{1}{\sqrt{2}} & \frac{1}{\sqrt{2}} \\ 1 & 0 & 0 \\ 0 & \frac{1}{\sqrt{2}} & -\frac{1}{\sqrt{2}} \end{bmatrix}; \quad P^{-1} = \begin{bmatrix} 0 & 1 & 0 \\ \frac{1}{\sqrt{2}} & 0 & \frac{1}{\sqrt{2}} \\ \frac{1}{\sqrt{2}} & 0 & -\frac{1}{\sqrt{2}} \end{bmatrix}.$$

Then $P^{-1}AP = \mathrm{diag}\{-1, 1, 8\}$. The matrix $B = P\,\mathrm{diag}\{-1, 1, 2\}P^{-1}$ is then such that $B^3 = A$. A simple calculation gives

$$B = \begin{bmatrix} 1\frac{1}{2} & 0 & -\frac{1}{2} \\ 0 & -1 & 0 \\ -\frac{1}{2} & 0 & 1\frac{1}{2} \end{bmatrix}.$$

8.19 The matrix is symmetric and its eigenvalues are 1 (of algebraic multiplicity 2) and 4. Since these are strictly postive, A is positive definite.

8.20 By definition, f is self-adjoint and $\langle f(x) \mid x \rangle > 0$ for every $x \in V$; and similarly for g. Then $f + g$ is self-adjoint and $\langle (f + g)(x) \mid x \rangle = \langle f(x) \mid x \rangle + \langle g(x) \mid x \rangle > 0$. Hence $f + g$ is positive definite.

8.21 If fg is positive definite then it is self-adjoint. Since f, g are self-adjoint it follows that $fg = (fg)^\star = g^\star f^\star = gf$.

Conversely, suppose that f, g are positive definite and commute. Then

$$\langle gf(x) \mid x \rangle = \langle fg(x) \mid x \rangle$$

where the left hand side is $\langle f(x) \mid g(x) \rangle$ and the right hand side is $\langle g(x) \mid f(x) \rangle$. It follows that $\langle gf(x) \mid x \rangle \in \mathbb{R}$ and so, by Theorem 8.12, gf is self-adjoint. Now by Theorem 8.15 we have $g = h^2$ where h is self-adjoint and, by Theorem 8.14, $h = q(g)$ for some polynomial q. Then f commutes with h. Consequently, for every non-zero $x \in V$, we have

$$\langle gf(x) \mid x \rangle = \langle h^2 f(x) \mid x \rangle = \langle hf(x) \mid h(x) \rangle = \langle fh(x) \mid h(x) \rangle > 0.$$

Hence gf is positive definite.

8.22 Let $\langle\!\langle - \mid - \rangle\!\rangle$ be an inner product on V. Then, by Exercise 7.23, for every $\beta \in V$ there exists a unique $\beta' \in V$ such that $\langle\!\langle - \mid \beta \rangle\!\rangle = \langle - \mid \beta' \rangle$. Define $f : V \to V$ by setting $f(\beta) = \beta'$. Then we have

$$\langle \alpha \mid f(\beta) \rangle = \langle\!\langle \alpha \mid \beta \rangle\!\rangle = \overline{\langle\!\langle \beta \mid \alpha \rangle\!\rangle} = \overline{\langle \beta \mid f(\alpha) \rangle} = \langle f(\alpha) \mid \beta \rangle.$$

Thus f is self-adjoint. Moreover,

$$\langle f(\alpha) \mid \alpha \rangle = \langle \alpha \mid f(\alpha) \rangle = \langle\!\langle \alpha \mid \alpha \rangle\!\rangle > 0$$

and so f is positive definite. As to the uniqueness of f, if g is also such that

$$(\forall \alpha, \beta) \qquad \langle\!\langle \alpha \mid \beta \rangle\!\rangle = \langle g(\alpha) \mid \beta \rangle$$

then clearly $g = f$.

9.1 (1) is bilinear; (2) is not.

9.2 The required matrices are

$$(1) \quad \begin{bmatrix} 2 & 0 & -3 \\ 0 & 2 & -5 \\ 4 & 0 & 0 \end{bmatrix}; \quad (2) \quad \begin{bmatrix} 3 & 0 & 0 \\ 0 & 2 & 0 \\ 0 & 0 & 1 \end{bmatrix}.$$

9.3 $-x_1 y_1 + -x_1 y_2 - 11 x_1 y_3 - 5 x_2 y_1 + 2 x_2 y_2 - 11 x_2 y_3 + 5 x_3 y_1 + x_3 y_2 - x_3 y_3$.

9.4 If A is symmetric and if there is an invertible matrix P such that $B = P^t A P$ then $B^t = P^t A^t P = P^t A P = B$.

9.5 $f(x, y) = p(x, y) + q(x, y)$ where $p(x, y) = \frac{1}{2}[f(x, y) + f(y, x)]$ is symmetric and $q(x, y) = \frac{1}{2}[f(x, y) - f(y, x)]$ is skew-symmetric. If also $f = a + b$ where a is symmetric and b is skew-symmetric then $a - p = q - b$ where the left hand side is symmetric and the right hand side is skew-symmetric, whence each is 0 and then $a = p$, $b = q$.

(1) $p = \frac{1}{2}[2x_1 y_1 + 3x_1 y_2 + 3x_2 y_1 + 2x_2 y_2]$ and $q = \frac{1}{2}[-x_1 y_2 + x_2 y_1]$.

(2) $p = f$ and $q = 0$.

9.6 The set of quadratic forms is closed under addition and multiplication by scalars.

9.7 $Q(x_1, \ldots, x_n) = \sum_{i=1}^{n} x_i^2$.

9.8 (1) The form is $\mathbf{x}^t A \mathbf{x}$ where

$$A = \begin{bmatrix} 3 & -\frac{5}{2} \\ -\frac{5}{2} & -7 \end{bmatrix}.$$

(2) The form is $\mathbf{x}^t A \mathbf{x}$ where

$$A = \begin{bmatrix} 3 & -\frac{7}{2} & \frac{5}{2} \\ -\frac{7}{2} & 4 & -2 \\ \frac{5}{2} & -2 & -3 \end{bmatrix}.$$

9.9 Since λ is an eigenvalue of A there exist $a_1, \ldots, a_n$ not all zero such that $A\mathbf{x} = \lambda\mathbf{x}$ where $\mathbf{x} = [a_1 \cdots a_n]'$. Then

$$Q = \mathbf{x}'A\mathbf{x} = \lambda\mathbf{x}'\mathbf{x} = \lambda\sum_{i=1}^{n} a_i^2.$$

9.10 We have

$$Q(x, y, z) = x^2 + 2xy + y^2 - 2xz - z^2 = (x + y)^2 + x^2 - (x + z)^2$$

so the required matrix is

$$\begin{bmatrix} 1 & 0 & 0 \\ 0 & 1 & 0 \\ 0 & 0 & -1 \end{bmatrix}.$$

9.11 (1) We have

$$2y^2 - z^2 + xy + xz = 2(y + \tfrac{1}{4}x)^2 - \tfrac{1}{8}x^2 + xz - z^2$$
$$= 2(y + \tfrac{1}{4}x)^2 - \tfrac{1}{8}(x - 4z)^2 + z^2$$

and so the rank is 3 and the signature is 1.

(2) Put $x = X + Y$, $y = X - Y$, $z = Z$ to obtain

$$2xy - xz - yz = 2(X^2 - Y^2) - (X + Y)Z - (X - Y)Z$$
$$= 2X^2 - 2Y^2 - 2XZ$$
$$= 2(X - \tfrac{1}{2}Z)^2 - \tfrac{1}{2}Z^2 - 2Y^2.$$

Thus the rank is 3 and the signature is -1.

(3) Put $x = X + Y$, $y = X - Y$, $z = Z$, $t = T$ to obtain

$$yz + xz + xy + xt + yt + zt = X^2 - Y^2 + 2XZ + 2XT + ZT$$
$$= (X + Z + T)^2 - Y^2 - Z^2 - T^2 - ZT$$
$$= (X + Z + T)^2 - (T + \tfrac{1}{2}Z)^2 - \tfrac{3}{4}Z^2 - Y^2.$$

Thus the rank is 4 and the signature is -2.

9.12 (1) The matrix in question is

$$A = \begin{bmatrix} 1 & -1 & 2 \\ -1 & 2 & -3 \\ 2 & -3 & 9 \end{bmatrix}.$$

Now

$$x^2 + 2y^2 + 9z^2 - 2xy + 4xz - 6yz = (x - y + 2z)^2 + y^2 + 5z^2 - 2yz$$
$$= (x - y + 2z)^2 + (y - z)^2 + 4z^2$$
$$= \xi^2 + \eta^2 + \zeta^2$$

where $\xi = x - y + 2z$, $\eta = y - z$, $\zeta = 2z$. Then

$$\begin{bmatrix} x \\ y \\ z \end{bmatrix} = P\begin{bmatrix} \xi \\ \eta \\ \zeta \end{bmatrix} = \begin{bmatrix} 1 & 1 & -\tfrac{1}{2} \\ 0 & 1 & \tfrac{1}{2} \\ 0 & 0 & \tfrac{1}{2} \end{bmatrix}\begin{bmatrix} \xi \\ \eta \\ \zeta \end{bmatrix}$$

and $P'AP = \text{diag}\{1, 1, 1\}$.

(2) Here the matrix is

$$A = \begin{bmatrix} 1 & 1 & 0 & -1 \\ 1 & 4 & 3 & -4 \\ 0 & 3 & 1 & -7 \\ -1 & -4 & -7 & -4 \end{bmatrix}.$$

The quadratic form is

$$\begin{aligned}
Q(x,y,z,t) &= (x+y-t)^2 + 3y^2 + z^2 - 5t^2 + 6yz - 6yt - 14zt \\
&= (x+y-t)^2 + 3(y+z-t)^2 - 2z^2 - 8t^2 - 8zt \\
&= (x+y-t)^2 + 3(y+z-t)^2 - 2(z+2t)^2 \\
&= \xi^2 + \eta^2 - \zeta^2
\end{aligned}$$

where $\xi = x+y-t$, $\eta = \sqrt{3}(y+z-t)$, $\zeta = \sqrt{2}(z+2t)$. Writing $\tau = t$, we then have

$$\begin{bmatrix} x \\ y \\ z \\ t \end{bmatrix} = P \begin{bmatrix} \xi \\ \eta \\ \zeta \\ \tau \end{bmatrix} = \begin{bmatrix} 1 & -\frac{1}{\sqrt{3}} & \frac{1}{\sqrt{2}} & -2 \\ 0 & \frac{1}{\sqrt{3}} & -\frac{1}{\sqrt{2}} & 3 \\ 0 & 0 & \frac{1}{\sqrt{2}} & -2 \\ 0 & 0 & 0 & 1 \end{bmatrix} \begin{bmatrix} \xi \\ \eta \\ \zeta \\ \tau \end{bmatrix}$$

and $P'AP = \text{diag}\{1,1,-1,0\}$.

9.13 The matrix is that of Exercise 9.10. Since the rank is 3 and the signature is 1 the quadratic form is not positive definite.

9.14 (1) $x^2 + 3xy + y^2 = (x + \frac{3}{2}y)^2 - \frac{5}{4}y^2$ so the rank is 2 and the signature is 0. Hence the form is not positive definite.

(2) $2x^2 - 4xy + 3y^2 - z^2 = 2(x-y)^2 + y^2 - z^2$ so the rank is 3 and the signature is 1. Hence the form is not positive definite.

9.15 Follow the hint.

9.16 We have

$$\begin{aligned}
\sum_{r,s=1}^{n} (\lambda rs + r + s)x_r x_s &= \lambda \sum_{r,s=1}^{n} (rx_r)(sx_s) + \sum_{r,s=1}^{n} (rx_r)x_s + \sum_{r,s=1}^{n} x_r(sx_s) \\
&= \lambda \Big(\sum_{r=1}^{n} rx_r \Big)^2 + 2 \Big(\sum_{r=1}^{n} x_r \Big) \Big(\sum_{r=1}^{n} rx_r \Big).
\end{aligned}$$

Now let

$$y_1 = \sum_{r=1}^{n} rx_r, \quad y_2 = \sum_{r=1}^{n} x_r, \quad y_3 = x_3, \quad \dots, \quad y_n = x_n.$$

Then the form is $\lambda y_1^2 + 2y_1y_2$, which can be written as

$$\begin{cases} \lambda(y_1 + \frac{1}{\lambda}y_2)^2 - \frac{1}{\lambda}y_2^2 & \text{if } \lambda \neq 0; \\ \frac{1}{2}(y_1 + y_2)^2 - \frac{1}{2}(y_1 - y_2)^2 & \text{if } \lambda = 0. \end{cases}$$

Hence in either case the rank is 2 and the signature is 0.

10.1 Let f^* be the adjoint of f. Then since we are dealing with a real inner product space we have, for all $x \in V$,

$$\langle f(x) \mid x \rangle = \langle x \mid f^*(x) \rangle = \langle f^*(x) \mid x \rangle.$$

It follows that

$$\langle f(x) \mid x \rangle = \langle \tfrac{1}{2}(f + f^*)(x) \mid x \rangle$$

where $\frac{1}{2}(f + f^*)$ is self-adjoint. If now $\langle f(x) | x \rangle = \langle g(x) | x \rangle$ where g is self-adjoint then we have $\langle (f - g)(x) | x \rangle = 0$ whence, by Theorem 10.3, $f - g$ is skew-symmetric. Consequently, since g is self-adjoint,

$$f - g = -(f - g)^* = -f^* + g^* = -f^* + g$$

and from this we obtain $g = \frac{1}{2}(f + f^*)$.

10.2 $g(-A) = g(A') = [g(A)]' = 0' = 0$.

Let the minimum polynomial of A be

$$m_A = a_0 + a_1 X + a_2 X^2 + \cdots + a_{n-1} X^{n-1} + X^n.$$

Since, by the first part of the question, $m_A(-A) = 0$ we have that

$$a_0 - a_1 X + a_2 X^2 - \cdots + (-1)^n X^n$$

is also the minimum polynomial of A. Hence $a_1 = a_3 = \cdots = 0$.

10.3 Compare with Example 10.2. The minimum polynomial of A is $X(X^2 + 9)$.

10.4 By Exercise 10.2, both A and $A' = -A$ have the same minimum polynomial. Then by the Corollary to Theorem 10.5, and Theorem 10.8, both A and A' have the same canonical form under orthogonal similarity.

11.1 The MAPLE code is

```
> m:=Matrix(4,4,(i,j)->4*(i-1)+j);
```

11.2 The MAPLE code is

```
> a:=Matrix([[1,2,3,4,5,6],[9,8,7,7,8,9],[1,3,5,7,2,4],
[5,4,5,6,5,6],[2,9,2,7,2,9],[3,6,4,5,6,7]]);
```

followed by

```
> ia:=a^(-1);
```

11.3 The MAPLE code is

```
> a:=Matrix([[x,2,0,3],[1,2,3,3],[1,0,1,1],[1,1,1,3]]);
```

followed by

```
> solve(Determinant(a)=0,x);
```

The answer is that the matrix is invertible except when $x = -\frac{1}{2}$.

11.4 Input A and B then:

```
> C:=A^3-B^2+A.B.(B^2-A);
```

followed by

```
> CharacteristicPolynomial(C,X);
```

11.5 The MAPLE code involves a loop:

```
> big:=-100
for a from 1 to 2 do
for b from 1 to 2 do
for c from 1 to 2 do
for i from 1 to 2 do
for j from 1 to 2 do
for k from 1 to 2 do
for x from 1 to 2 do
for y from 1 to 2 do
for z from 1 to 2 do
m:=Matrix(3,3,[[a,b,c],[i,j,k],[x,y,z]]);
if Determinant(m)>big then big:=Determinant(m) fi;
od;od;od;od;od;od;od;od;od:
print(big);
```

The output is 5.

For the second part, the code is

```
> for a from 1 to 2 do
for b from 1 to 2 do
for c from 1 to 2 do
for i from 1 to 2 do
for j from 1 to 2 do
for k from 1 to 2 do
for x from 1 to 2 do
for y from 1 to 2 do
for z from 1 to 2 do
m:=Matrix(3,3,[[a,b,c],[i,j,k],[x,y,z]]);
if Determinant(m)=big then print(m) fi;
od;od;od;od;od;od;od;od;od:
```

The answer is

$$
\begin{bmatrix} 1 & 2 & 2 \\ 2 & 1 & 2 \\ 2 & 2 & 1 \end{bmatrix}, \quad
\begin{bmatrix} 2 & 1 & 2 \\ 2 & 2 & 1 \\ 1 & 2 & 2 \end{bmatrix}, \quad
\begin{bmatrix} 2 & 2 & 1 \\ 1 & 2 & 2 \\ 2 & 1 & 2 \end{bmatrix}.
$$

11.6 The MAPLE codes are the following:

```
> f:=(i,j)->i^2+j^2;
```

```
> for i from 1 to 10 do
A:=Matrix(i,i,f);
Determinant(A);
od;
```

```
> g:=proc(i,j);
if i<j then i^2+j else j^2 fi;
end:
```

```
> for i from 1 to 10 do
A:=Matrix(i,i,g):
B:=Matrix(i,i,f);
C:=A-B;
print(Determinant(A),Determinant(C),
Determinant(C)/Determinant(A));
od:
```

```
> la:=[seq(Determinant(Matrix(i,i,g)),i=1..20)];
```

```
> lb:=[seq(Determinant(Matrix(i,i,g)-Matrix(i,i,f)),i=1..20)];
```

```
> lc:=[seq(i^2+lb[i]/la[i],i=1..20)];
```

11.7 The conjecture is that if $A = [a_{ij}]_{n \times n}$ is such that a_{ij} is a polynomial in i and j then, except for some small values of n, the determinant of A is 0.

Index